Is Alcoholism Hereditary?

ALSO BY DONALD W. GOODWIN

Anxiety

Alcohol and the Writer: an American Epidemic

Psychiatric Diagnosis, fourth edition,
Co-authored with S. B. Guze

Phobia: The Facts

Longitudinal Research in Alcoholism
Co-authored with K. T. Van Dusen
and S. A. Mednick

Alcoholism: The Facts

Alcoholism and Affective Disorders
Co-edited with Carlton Erickson

Is Alcoholism Hereditary?
(First edition)

Is Alcoholism Hereditary?

SECOND EDITION

Donald W. Goodwin, M.D.

BALLANTINE BOOKS • NEW YORK

 Published in the United States by Ballantine Books, a division of Random House, Inc., New York, and simultaneously in Canada by Random House of Canada Limited, Toronto. A previous edition of this book was published by Oxford University Press in 1976.

Grateful acknowledgment is made to the following for permission to reprint previously published material:

Year Book Medical Publishers, Inc.: graph from *Alcoholic Beverages in Clinical Medicine* by C.D. Leake and M. Silverman. Copyright © 1966 by Year Book Medical Publishers, Inc., Chicago. Used by permission.

Sterling Lord Literistic, Inc. and Jonathan Cape Ltd.: excerpts from *Lunar Caustic* by Malcolm Lowry. Copyright © 1963 by Margerie Lowry. Reprinted by permission of the Executors of the Malcolm Lowry Estate and Sterling Lord Literistic, Inc.

Viking Penguin, Inc.: excerpts from *The Medusa and the Snail* by Lewis Thomas. Copyright © 1977 by Lewis Thomas. Reprinted by permission of Viking Penguin, Inc.

Trust of Irving Shepard: excerpts from *John Barleycorn* by Jack London. Permission granted by I. Milo Shepard on behalf of the Trust of Irving Shepard "B."

Library of Congress Catalog Card Number: 87-91549

ISBN: 0-345-34821-4

Cover design by James R. Harris
Book design by Beth Tondreau Design

Manufactured in the United States of America
Second Edition: March 1988

TO BILL GOODWIN
(BOTH OF THEM)

Contents

Acknowledgments

As always, Mrs. Evelyne Karson did a superb job of preparing the manuscript while simultaneously doing a trillion other things. I am most grateful.

Is Alcoholism Hereditary?

Introduction

When the original edition of *Is Alcoholism Hereditary?* was published in 1976, little was known about possible genetic factors in alcoholism. Studies had shown that alcoholism ran in families. In fact, a Danish study had just shown that alcoholism ran in families even when the sons were separated from the alcoholic parents and raised by nonalcoholic adoptive parents. (Nothing yet was known about the adopted daughters.)

Other research results were ambiguous. One study of twins pointed to a genetic factor in alcoholism, but another did not. Social scientists recognized that some features of alcoholism were hard to attribute to genes, for example, the low rate of alcoholism in Jews, the high rate in the Irish. They suspected that it was tough on children being raised by alcoholic parents, but nobody had started studying these children. If some of the children became alcoholic, it was understandable. Their home lives could be horrendous. But how could scientists account for that adoption study in Denmark that showed that sons of alcoholics became alcoholic without a horrendous home life?

Since the early 1930s, hereditary studies had not been popular. Hitler had given heredity a bad name with all the talk about superior races. One of America's most distinguished psychiatrists, Adolph Meyer, the Freudians and neo-Freudians, and the new class of social workers all stressed the importance of environmental factors on behavior. This, after all, seemed the optimistic view to take. If you had somehow "learned" to become neurotic or alcoholic—or even schizophrenic—maybe you could unlearn it, with the help of a little tinkering with the unconscious, some counseling, or a better job. Whereas if you had been born that way, treatment seemed futile.

In the early 1970s, the opinion of social scientists was virtually unanimous: Blue eyes were inherited, alcoholism was not.

Then, in less than a decade, came a dramatic change. First, researchers in St. Louis reported that sons of alcoholics from broken homes were six times more likely to become alcoholic if the biological parent had been alcoholic than if the custodial father was alcoholic. The first Danish adoption study came out almost simultaneously and reported essentially the same thing.

These reports were followed, in fast order, by two other adoption studies—one in Sweden, one in Iowa. They also found the same thing: Susceptibility to alcoholism ran in families whether the children of alcoholics were raised by the alcoholic families or not. For a time this seemed more demonstrable with men than women, but recent studies also indicate that women separated from their alcoholic biological parents are equally vulnerable to alcoholism.

The adoption studies awakened a new interest in possible hereditary factors in alcoholism. Scores of investigators sought funds to support genetic studies in alcoholism. The studies went on simultaneously in several countries and involved scientists from many disciplines. Psychiatrists and sociologists began comparing alcoholism that ran in families with alcoholism that did not, and a concept called "familial alcoholism" was developed.

By 1980, at least a dozen groups of investigators were studying the children of alcoholics—nonadopted as well as adopted—and comparing them with children of nonalcoholics. There were tantalizing findings:

1. Children of alcoholics have a higher tolerance for alcohol than do children of nonalcoholics.
2. Children of alcoholics underrespond to certain stimuli recorded on electroencephalograms (EEGs).

3. Children of alcoholics generate more alpha activity on the EEG.
4. Children of alcoholics are more often hyperactive than other children.

Studies comparing children of alcoholics with other children are the first phase of long-term studies to identify factors associated with future development of alcoholism. They are called high-risk studies because children of alcoholics are at high risk for becoming alcoholic. Based on numerous family studies, one can anticipate that one of every four or five male offspring of alcoholics will become alcoholic. Large amounts of information are now being gathered about these children, and the studies will yield helpful data as these children mature into adults. Nobody can anticipate what types of information obtained now will predict future alcoholism, but out of the mass of information obtained—biological, psychological, and sociological—predictors surely will emerge.

Meanwhile, scientists are pondering another question. If alcoholism is inheritable, *what* is being inherited? There are some interesting leads. With sufficient funding over the next few years, chemical mechanisms underlying alcohol dependence may be identified. Once this happens, altering these mechanisms in the interest of preventing addiction will become realizable.

At the same time, treatment facilities for alcoholism have mushroomed, and their treatment programs are influenced by new information about familial and genetic factors, e.g., the growing evidence that familial alcoholics differ in important aspects from nonfamilial alcoholics. There is some evidence that nonfamilial alcoholics more often have psychiatric disorders, including depression and phobias.

The plight of children raised by alcoholics has achieved international recognition and is the subject of numerous conferences and books. All this has led to humane and informed action to alleviate the situation of these victims. The National Institute of Alcoholism and Alcohol Abuse has identified family and genetic research as top priorities for research funding.

The most promising development of all of the past ten years is the growing number of highly trained and skilled scientists at all levels who are now involved in alcoholism research. Books and scientific journals on biological aspects of alcoholism have multiplied, and hardly a month goes by without a conference somewhere in the world at which scientists report their latest findings about alcoholism and genetics. Alcoholism, a disease once considered incurable, has become a respectable subject for scientific inquiry, and the stigma that once branded the disease as well as those who studied it is rapidly disappearing.

This thoroughly revised edition of *Is Alcoholism Hereditary?* takes a close look at these new developments. I have expanded my original book to include up-to-date information on alcohol metabolism, alcoholism, genetics, adoption and twin studies, familial alcoholism, causation, and treatment.

I start with the assumption that the extent of the reader's knowledge about alcohol is that you sometimes rub yourself with it and sometimes drink it, but shouldn't do both from the same bottle. So the book begins at the beginning, with yeast, and discusses alcoholic beverages, what happens to alcohol in the human body, and why it affects people differently. Unless you know something about alcohol, you will not be in a position to understand much about alcoholism, the subject of chapter 2.

Chapter 2 opens with an alcoholic's own story, a story with a moral: Alcoholism is no joking matter. We have a tendency to joke about it (no one would dream of joking about cancer), and no doubt there are psychological reasons for this; alcoholics do it themselves. But the utter seriousness of alcoholism can't be questioned.

Alcoholism touches all our lives. Apart from their own problems, most alcoholics have families that suffer with them, and employers who must endure or fire them. Many drive cars that, with statistically predictable regularity, strike other cars. To reduce our population of alcoholics by even a fraction would make life a good deal more pleasant for a large number of people.

The second chapter also deals with the definition of alcoholism; its symptoms and progression; why some people become alcoholics; and questions that interest anyone who drinks, such as: Does alcohol harm the brain or liver?

The groundwork is now laid for the question raised again and again in this book: *Why* does alcoholism run in families? Chapter 3 gives the evidence that it does run in families, and convincing evidence it is.

Not everything that runs in families is hereditary, and chapter 4 deals with strategies for teasing apart the contributions of heredity and environment in the causation of family illnesses.

Chapter 5 shows how these strategies have been applied to alcoholism. This chapter and the next summarize the facts and arguments for and against the importance of heredity in alcoholism.

Where do we go from here? Chapter 7 discusses how some of the big questions may be answered someday and suggests some things that might be accomplished in the meantime.

Chapter 8 reviews what resources are now available to help the alcoholic. This chapter is often critical, but there is no choice, if you respect evidence. I have my own method of treating alcoholism, and it seems to work, at least for a time. I describe it in some detail. The chapter concludes with suggestions for doctors and families who must confront the problem of an alcoholic.

Finally, there are areas I know are controversial but about which I have to take a stand, if only because too much hedging tries a reader's patience. Some of the hedges are in the Notes and References section, together with information about further reading.

D.W.G.

CHAPTER ONE

Alcohol

You have asked me how I feel about whisky. All right, here is just how I stand on this question:

If, when you say whisky, you mean the devil's brew, the poison scourge, the bloody monster that defiles innocence, yea, literally takes the bread from the mouths of little children; if you mean the evil drink that topples the Christian man and woman from the pinnacles of righteous, gracious living into the bottomless pit of degradation and despair, shame and helplessness, then certainly I am against it with all of my power.

But, if you when you say whisky, you mean the oil of conversation, the philosophic wine, the stuff that is consumed when good fellows get together, that puts a song in their hearts and laughter on their lips and the warm glow of contentment in their eyes; if you mean Christmas cheer; if you mean the stimulating drink that puts the spring in the old gentleman's step on a frosty morning, if you mean the drink that enables a man to magnify his joy, and his happiness, and to forget, if only for a little while, life's great tragedies and heartbreaks and sorrows, if you mean that drink, the sale of which pours into our treasuries untold millions of dollars, which are used to provide tender care for our little crippled children, our blind, our deaf, our dumb, our pitiful aged and infirm, to build highways, hospitals, and schools, then certainly I am in favor of it.

This is my stand. I will not retreat from it; I will not compromise.

—Address to the legislature
by a Mississippi state
senator in 1958

Yeast is alcohol's first victim. This is only poetic justice, since yeast started the problem in the first place.

When yeast grows in sugar solutions without air, most of the sugar is converted (fermented) into carbon dioxide and alcohol. Carbon dioxide is what makes liquids bubble ("fermentation" comes from the Latin word for "boil") and champagne corks pop. Yeast excretes alcohol as it ferments. Most drinkers do not know they are drinking yeast excrement. Would it matter?

It matters for yeast. When the alcohol concentration reaches about 12 or 13 percent, the yeast dies of acute alcohol intoxication. This is why most unfortified wines, produced by fermentation alone, have alcohol concentrations of no more than 12 or 13 percent. Sherry, port, and other fortified wines have alcohol added.

As a rule, people do not drink pure alcohol. They drink alcoholic beverages which are mostly water and ethyl alcohol, a two-carbon straight-chain alcohol that is really just water with an ethyl group substituting for a hydrogen. However, tiny amounts of other chemicals are present in alcoholic beverages that provide most of the taste and smell, and all of the color, if any. These other chemicals, in toto, are called congeners. Because of congeners, beer can be distinguished from brandy, although both consist almost entirely of ethyl alcohol and water. Depending on the beverage, congeners include varying amounts of amino acids, minerals, and vitamins; a one-carbon alcohol called methanol or "wood alcohol"; and the "higher" alcohols with more than two carbons, otherwise known as fusel oil.

Even in small quantities, wood alcohol and fusel oil are poisons. So is ethyl alcohol, but a lot more of it is required to do damage. Is there enough wood alcohol and fusel oil in a highball to hurt anyone? Probably not, but no one is sure. To be on the safe side, some people avoid drinks with the most congeners—whiskey and brandies—and drink relatively congener-free vodka. They are not aware that vodka often contains more wood alcohol—notorious for causing blindness in excess—than other beverages. Although there is almost certainly not enough wood alcohol in vodka to cause blindness, it does not improve vision either.

Apart from man's contribution—the brewer's art, the cosseted grape—beverages differ according to the sugar source. From grapes, wine; from grain and hops, beer; from grain and corn, whiskey; from sugarcane, rum; and originally from the lowly potato, but now mainly from grain, vodka.

Man's great achievement in improving upon yeast's modest productivity was distillation, discovered about A.D. 800 in Arabia (*alcohol* comes from the Arabic *alkuhl,* meaning "essence"). Distillation boils away alcohol from its sugar bath and collects it as virtually pure alcohol. Then, because pure alcohol is pure torture to drink, it is diluted with water, so that instead of having 100 percent alcohol, you have 50 percent or 100-proof alcohol (percent being half the proof).

ALCOHOL'S FATE IN THE BODY

What happens to alcohol when you drink it? Essentially the same thing that happens if you don't drink it. It turns to vinegar.

When alcohol "sours" in the open air, bacteria are responsible. To become vinegar (acetic acid) in the body, alcohol needs two enzymes: alcohol dehydrogenase and aldehyde dehydrogenase. The first is located in the liver in a surprisingly large supply. Surprising because, as far as we know, alcohol dehydrogenase does nothing except metabolize alcohol. It is there in all mammalian livers—in the horse's in particular plentitude. Why? Did Mother Nature anticipate that someday a mammal like man would develop a taste for alcohol and need a way to dispose of it? Or did it happen that millennia ago, horses and other vegetarians ate fermenting fruit lying on the ground, and their obliging livers evolved a helpful enzyme?

Nobody knows, but it is a handy enzyme to have in any case. It disposes of 100-proof distilled spirits at the rate of about one ounce per hour, slow enough to soak the brain without, one hopes, pickling it.

As alcohol breaks down into acetic acid it first becomes an intermediary chemical, an aldehyde, which is very toxic. However, the second enzyme, aldehyde dehydrogenase, which is found not just in the liver but throughout the body, quickly turns the aldehyde into harmless acetic acid. Fed into the body's normal metabolic machinery, acetic acid becomes carbon dioxide and water, burning or storing seven calories per gram of alcohol in the process.

HOW ALCOHOL AFFECTS THE BODY

Vinegar is harmless, but the process that produces it may not be. As it is oxidized, alcohol is progressively stripped of hydrogen atoms, which results in some interesting biochemical changes that may or may not be harmless (the evidence is not in yet). Some of the changes are:

1. There is an increase in lactic acid. This is interesting because a connection between increased lactic acid and anxiety attacks has been observed, and heavy drinking is also associated with anxiety attacks.
2. There is an increase in uric acid. This is interesting because increased uric acid is associated with gout, and for centuries gout has been associated with alcohol.
3. There is an increase in fat—not the slow increase that comes from calories (those seven calories per gram) but a rapid increase from the oxidation of alcohol. This fat is seen mainly in the liver or blood. One night of serious drinking—say, six or seven highballs—discernibly increases the fat content of the liver. The liver will be fattier still if the drinker also consumes fatty food.

Is a fatty liver bad? The connection between a fatty liver and liver diseases, such as hepatitis and cirrhosis, is unclear. For one thing, the fat goes away soon after the drinker stops drinking. Also, most people drink but most do not develop liver disease. Among those very heavy drinkers we call alcoholics, perhaps only 5 or 10 percent develop liver disease, although presumably their livers are fatty most of the time. On the other hand, people who develop a particular type of liver disease called Laennec's cirrhosis usually are heavy drinkers.

Many disorders associated with heavy drinking are apparently caused by malnutrition, but this may not be true of cirrhosis. Laennec's cirrhosis has been produced in well-nourished baboons after four years of drunkenness. Most of the drunken baboons, however, only had fatty livers, and controversy still thrives about whether alcohol alone causes cirrhosis.

Intoxicating amounts of alcohol also increase levels of fat in the bloodstream. In high enough doses, particularly when combined with a fatty meal, alcohol may even produce visible fat in the blood. The plasma takes on a faint milky tinge, probably a more frightening development for most people than the notion of having fat in their liver. Still, while it does not sound good, nobody knows how bad it is, and possibly it is not harmful at all.

These effects of alcohol on the body have been known for many years. In the past decade there has been increased effort by biochemists to understand the chemical elements of intoxication and addiction. Here are four recent discoveries that have opened up new avenues of exploration to the problem of alcoholism.

Membrane studies

The body is made up of cells, and every cell is surrounded by a membrane. Within the cells are other structures also surrounded by membranes. Research into these membranes' structure and function has provided new information about the effects of alcohol on the body.

Biological membranes are largely composed of fat in a semisolid state. Alcohol increases the fluidity of membranes, producing alterations in their many functions, such as controlling the transport of compounds in and out of the cells.

Alcohol has been shown to cause membranes to become more fluid initially, but chronic exposure of membranes to alcohol makes them more rigid, apparently to compensate for the preceding increase in fluidity and disturbance of function. It has been suggested that this change in membrane fluidity corresponds to tolerance (the ability of the body to adapt to increasing amounts of alcohol). Withdrawal symptoms, which occur when a person stops drinking heavily, may in turn reflect these membrane alterations. This work is still at an early stage but illustrates how alcohol research has advanced from behavior to organs to the molecular level.

TIQs

It has been discovered that small amounts of alcohol produce compounds in the brain closely resembling morphine. These compounds are called tetrahydroisoquinalones (TIQs), which are formed in minute quantities but are biologically active. When injected into the brains of animal subjects, TIQs cause the animals to drink more than usual. The increased drinking persists even after the TIQs are no longer injected. In many ways, TIQs behave like morphine. They combine with certain morphine (opiate) receptors in the brain. Drugs that block the effects of morphine also block certain effects of alcohol, such as incoordination and coma. There is also evidence that alcoholics experiencing alcohol withdrawal apparently have increased levels of TIQs in their spinal fluid.

None of this means that TIQs are directly related to the cause of alcoholism; relevant studies have yet to be done. One such study might involve giving alcohol to sons of alcoholics and sons of nonalcoholics. As a group, sons of alcoholics have a risk of becoming alcoholic about four or five times greater than that of sons of nonalcoholics. As will later be explained, this predisposition to alcoholism in the children of alcoholics may involve genetic factors. If sons of alcoholics respond to alcohol with higher levels of TIQs than sons of nonalcoholics, it would be strong evidence that TIQs indeed play a causal role in alcoholism.

Neurotransmitter studies

Chemicals that transmit messages between nerve cells are called neurotransmitters. In recent years there have been numerous investigations into the possible role of neurotransmitters in alcoholism.

A neurotransmitter called serotonin has been most often studied. It seems that alcohol first elevates serotonin levels in the brain and then produces a depletion of serotonin. Animals drink more if their serotonin supplies are depleted. This and other evidence suggests that serotonin may play a role in producing the "highs" and "lows" associated with drinking and thus may be a factor in alcohol addiction.

Another transmitter implicated in the action of alcohol is beta-endorphin. This compound occurs naturally in the brain and was discovered after it was found that the nervous system contains opiate receptors—proteins on cell membranes that bind specifically with morphine-like compounds called opiates. There is evidence that alcohol increases endorphin levels and then subsequently leads to a reduction of these levels. There has been speculation that alcoholics are "born" with subnormal levels of endorphins and drink in order to correct this deficiency. Without direct evidence to support the theory, it nevertheless has resulted in the marketing of pills believed to increase endorphin levels and thus decrease the craving for alcohol. The efficacy of the treatment remains to be shown.

Benzodiazepine studies

In the early 1960s a new class of anxiety-relieving chemicals called benzodiazepines was introduced on the market. The most popular types were Librium and Valium. There are now perhaps a dozen benzodiazepine-type drugs available, one of the most recent being Xanax.

These drugs are commonly used to relieve withdrawal

symptoms after drinking. Their success in this regard has led to speculation that perhaps alcohol involves chemical reactions similar to that produced by the drugs. Just as the brain has receptors for opiates, it also has receptors for the benzodiazepine drugs. Alcohol alters the activity of these receptors. Chronic exposure to alcohol in mice leads to a reduction in the number of receptors and this has led to a hypothesis: People predisposed to alcoholism may be born with a deficiency of benzodiazepine receptors, or chronic drinking may reduce the number of such receptors. People may drink to compensate for the deficiency and overcome anxiety, but relief is only temporary as the receptors become saturated and even more deficient. This creates the self-perpetuating "addictive cycle" described in chapter 2. No specific receptors for alcohol have been found and the story, at this writing, is still unfolding.

HOW ALCOHOL AFFECTS BEHAVIOR

The effects of any drug depend mainly on the dose. The chance of death occurring from a sip of beer is remote. However, a quart of whiskey drunk in an hour will kill most men. This dose–effect rule applies to any substance a person consumes. People die from drinking too little water, and from drinking too much. A little strychnine may even be good for you (it helps rats concentrate). But how do you determine the dose? Does alcohol refer to a bottle of 3.2 beer or to a quart of bourbon? The amount of alcohol consumed is the main factor in determining its effects, but there are other factors that must be considered as well.

Concentration of alcohol in the blood

What really counts is not how much alcohol a person drinks but how much gets into the bloodstream. This in turn depends on many things.

Some alcohol is absorbed through the stomach wall, but

most reaches the bloodstream through the small intestine. Between the stomach and small intestine is a muscular ring called the pyloric valve. When the valve clamps shut, as may happen when it is jolted by a straight shot of whiskey, the alcohol remains in the stomach, where it is absorbed at a very slow pace. For people with sensitive pyloric valves, a strong shot of whiskey or an extra-dry martini may be self-defeating if a quick effect is the goal.

For rapid absorption, it is important that the alcohol reach the small intestine in the highest possible concentration and the shortest possible time. People who have had their pyloric valves removed surgically, as in the treatment of ulcers, find they get drunk faster than previously. Without a pyloric valve to slow down its passage, alcohol quickly swooshes into the small intestine and thence into the bloodstream.

Other factors influencing absorption include the presence or absence of food in the stomach and the type of beverage. When alcohol has to compete with other, larger, and often more aggressive molecules in crossing the gastric and small intestinal membranes, it isn't as readily absorbed. In one experiment, alcohol in the form of different beverages was given to subjects with and without food. Although the same amount of alcohol was consumed over the same length of time, the subjects' blood–alcohol concentration varied greatly. Gin on an empty stomach produced a peak blood–alcohol level above the legal concentration for drivers in most states. Beer combined with a meal resulted in a peak blood level legally compatible with driving in any state. The graph on page 17 illustrates this.

There has been speculation about the difference in alcoholism rates between two wine-drinking countries, France and Italy. Italy, with the lower rate, has a national tradition of drinking wine mainly with meals, while the French tend to drink wine between meals as well as with them. When wine and spaghetti compete for transport across the intestinal wall, it is not surprising that spaghetti, finishing the race first, will prevent the wine from making much headway.

Mixing food and alcohol produces a slight increase in the

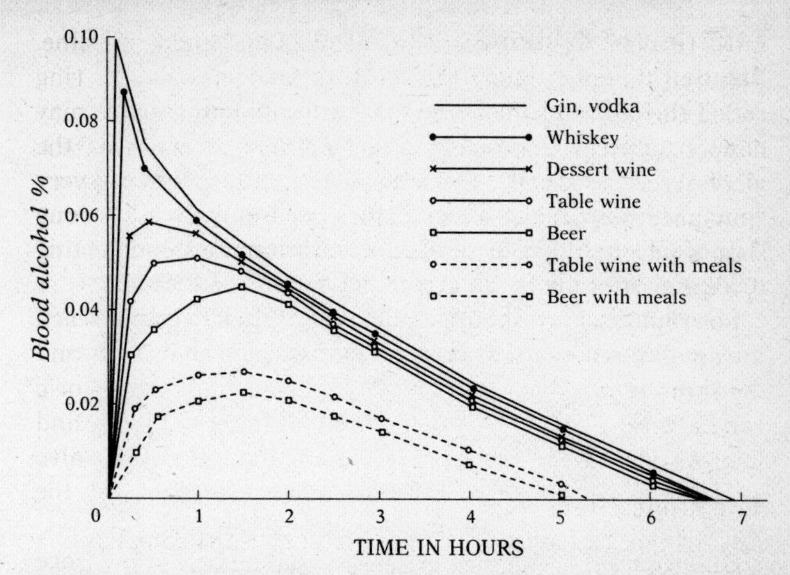

Typical blood-alcohol curves resulting from ingestion of various spirits, wines, and beer, each at amounts equivalent to 0.6 gm of alcohol per kilogram of body weight.

oxidation of alcohol and hence its removal from the bloodstream. This may partly explain the lower blood–alcohol level when food and alcohol are combined. It also may partly explain the known reluctance of alcoholics to eat while drinking, since presumably the alcoholic has no strong desire to remove alcohol from his bloodstream any faster than necessary.

Rate of absorption

The effect of alcohol also depends on how quickly it gets to the bloodstream. In general, the faster the rate of absorption, the more striking the effect. This may help explain the popularity of a dry martini before meals.

Duration of drinking

The body adapts rapidly to chemical insults, including alcohol. The longer alcohol remains in the blood, the more its effects diminish. In practical terms (for a secretary or writer, anyway), if you make five errors per minute while typing sober, you may make fifteen errors per minute while typing with a certain blood–alcohol concentration after one hour of drinking, but only seven errors per minute at the same concentration after five hours of drinking. After five hours you may not care how many errors you make, but that is another consideration.

The slope effect

Any drinker can tell you he feels better getting drunk than he does sobering up. That is, as the blood–alcohol level climbs from A to B to C, the drinker may feel euphoric at B and C, but as the blood–alcohol level falls from C to B to A, the drinker may not only feel no euphoria at B, but may actually feel discomfort, presaging the hangover to come at A. This "slope" effect is closely related to and hard to separate from the duration effect.

Tolerance

As people drink more over days, months, and years, they gradually *need* to drink more to obtain the same effect. This is called tolerance. Its importance is often exaggerated. A seasoned alcoholic at the prime of his drinking capacity may be able to drink, at most, twice as much as a teetotaler of similar age and health. Compared with tolerance for morphine, which can be manyfold, tolerance for alcohol is modest.

More striking than "acquired" tolerance may be inborn tolerance. Individuals vary widely in the amount of alcohol they can tolerate independent of drinking experience. Some people, however hard they try, cannot drink more than a small amount of alcohol without developing a headache, upset stomach, or dizziness. They rarely become alcoholic but deserve no credit for it. Their "alcohol problem" is that they cannot drink very much.

Others seem able to drink large amounts with hardly any bad effects. It appears they were born with this capacity and did not develop it entirely from practice. They can become alcoholic, and some do.

Differences in tolerance for alcohol apply not only to individuals but to racial groups. For example, some Oriental groups develop flushing of the skin, sometimes accompanied by nausea, after drinking only a little alcohol. For obvious reasons, alcoholism is rare in these groups. Native Americans are also believed to be intolerant of alcohol, but the nature of the intolerance is more ambiguous and does not appear to discourage heavy drinking.

Set and setting

Any drug response that involves thinking and mood is bound to be influenced by expectation. Alcohol is no exception. If a person believes alcohol will improve his mood, diminish fatigue, make him feel sexy, or have other salutary effects, the chances that these pleasant changes will occur may be improved. The same goes for expectations of unpleasant changes.

In medicine this is called the placebo effect: Drugs tend to do for people what they expect them to do. Expectation presumably has little or no role in the treatment of pneumonia with antibiotics—although one cannot be certain—but it has powerful implications for treating psychiatric disorders. Sometimes sugar pills help almost as much as expensive and potentially toxic tranquilizers or antidepressants.

It is the gap between the sugar pill's performance and that of the "active" drug that justifies the prescription of the drug, and sometimes the gap is quite small. Placebos can even produce side effects—headaches, nausea, rashes. Everyone is a little susceptible to suggestion, some more than others.

It is difficult to know how much the effect of alcohol in any given person on any particular occasion is influenced by expectations (what psychologists call "set"). Presumably, the stronger the dose, the smaller the placebo effect. But it is a common laboratory and social observation that some people get "drunk" on very little alcohol. This may be because they want or expect to get drunk quickly.

But set refers to more than expectation. If a person is tired, alcohol may have more of an anti-fatigue effect than usual. If he is hungry, it may make him more hungry (or less). If his mood is good, it may become better. If bad, worse. All of this refers to set—the psychological and physical state of the person at the time he proceeds to drink.

To a considerable extent, set is linked to setting. Where is the person drinking? With whom? If he enjoys the people he is with, he may also enjoy the alcohol more. If the occasion is a celebration, a drink may have a livelier effect than would the same amount taken routinely before dinner.

Alcohol is said to make people talk louder, and this often seems true. On the other hand, two men in a duck blind, nipping bourbon to warm up, may talk more softly than usual.

The importance of set and setting in shaping a person's response to alcohol should not be underestimated, although it is difficult to study their relative influence at a given time.

THE FOUR STAGES OF INTOXICATION

There is an old saying that alcohol affects a person in four ways. First, he becomes jocose, then bellicose, then lachrymose, and finally comatose.

Comatose he does indeed become if he drinks enough, but the other three stages are not inevitable. Some people hardly feel jocose at all. One reason may be that they do not want to feel jocose. Their reasons for drinking may be purely social; others drink, so must they.

Many people become argumentative when they drink and some become combative, but these responses are strongly influenced by social circumstances. The legendary barroom brawl usually occurs in lower-class bars and is a rarity in upper-class saloons. Countless parties are held nightly in middle-class suburbia and, although drinking is common, fighting is not.

This is not to deny that drinking may bring out the beast in man (as will barbiturates, amphetamines, and other drugs, including marijuana). Alcohol is involved in at least half of all homicides in this country, with either the attacker, the victim, or both under the weather. This probably explains why more murders occur on Saturday night than on any other evening (the fewest occur on Tuesdays). Again, the connection between alcohol and bellicosity has class overtones, since most murders occur in the lower and lower-middle classes. It has been suggested that one reason fights occur in barrooms is that rarely are so many people thrust together so closely for such long periods, with hardly anything to do but talk, drink, and fight.

One of the paradoxes about alcohol is that people sometimes cry when they drink. Why, then, drink? Isn't the whole point of drinking to feel happier? The fact is, even though some people become anxious and depressed when they drink, they do not give up drinking, which challenges the widely held assumption that people drink mainly to feel *less* anxious and depressed. The motives for drinking, in truth, are complex and inscrutable, with no single explanation sufficing for all circumstances.

Alcohol is often described as a depressant drug that depresses first the higher centers in the brain and then downwardly anesthetizes the brain until finally, in lethal dosage, it snuffs out life itself by depressing the respiratory center at the base of the brain. This, like most things said about alcohol, is an oversimplification.

What is alcohol "depressing"? Usually not activity. Most people get a lift from alcohol and many become more animated and active. Nerve fibers fire about as readily in an alcohol solution as they do otherwise unless the concentration is far above what most people can achieve by drinking. It is sometimes said that by depressing the higher centers of the brain, alcohol releases the lower centers and that this is why people are more uninhibited when they drink—the "animal instincts" are released. However, studies do not support this theory of the top-to-bottom action of alcohol. Coordination, a lower function, is often impaired at lower doses of alcohol than is memory, a higher function.

Again, dosage is crucial. Alcohol in rather small doses may improve certain types of performance. Apparently this is most likely to occur in activities where the person is not very proficient and where the effects of increased confidence might be expected to arise. If someone does poorly at hitting the target on a firing range, for example, he may improve somewhat after several drinks of alcohol. On the other hand, if he does well normally, his performance may fall off when he takes in small amounts of alcohol. Nevertheless, in moderate to high amounts, alcohol usually diminishes function across the board.

An interesting exception to this general rule has emerged in several recent studies. Apparently if a person learns certain things, such as word lists, while intoxicated—even severely intoxicated—he will remember them better when reintoxicated than when sober. Called "state-dependent learning," this is one of the few exceptions to the overall impairing effect of alcohol at moderate and high doses.

Alcohol does something else that is almost unique among drugs. It produces a classical amnesia called "blackout" in which the drinker does highly memorable things while drinking but cannot remember them the next day. Many social drinkers have had this experience, but it occurs most frequently in alcoholics and will be discussed in the next chapter.

ARE THERE BENEFICIAL EFFECTS FROM ALCOHOL?

There are two questions that cause controversy: (1) Are alcoholics able to return to controlled drinking? (2) Is alcohol, even in modest amounts, ever good for you? The first question will be dealt with, gingerly, in chapter 8. Meanwhile, a book on alcohol is not complete without addressing the possibility that some effects of alcohol may be beneficial. Here is the evidence for this:

Moderate drinkers seem to live longer than teetotalers or heavy drinkers

The news about heavy drinkers comes as no surprise, but the evidence raises a question about teetotalers: Why are teetotalers teetotalers? Is it possible they are abstainers because of health problems, such as diabetes or kidney disease? Health problems that discourage drinking may also shorten life expectancy, explaining the correlation between abstinence and reduced longevity. At least one study has addressed this issue and reports that abstainers were *not* abstaining because of health reasons, but the explanation cannot be totally discounted.

Drinking increases a form of cholesterol in the blood called high density lipoprotein (HDL)

HDL is sometimes called "good" cholesterol because high levels of HDL are associated with a decreased risk of coronary artery disease. Exercise also increases HDLs. One study concluded that two beers had the same effect on HDLs as a half hour of jogging. Another report points out that HDLs are found in several plasma fractions and that the fraction associated with decreased risk of heart attacks is not the fraction increased by drinking. It is, however, the fraction associated with heavy drinking. Nevertheless, while controversial, a growing body of literature suggests that alcohol, perhaps by elevating HDLs, reduces the risk of heart attack, and, following one heart attack, reduces the chance of recurrence.

Should doctors recommend alcohol for this reason? This is debatable. Heavy drinking is associated with at least three cardiovascular diseases: hypertension, enlargement of the left ventricle (reducing cardiac efficiency), and arrhythmias. The correlation between these conditions and drinking is not strong, but it does exist. Therefore, a trade-off is involved. Drinkers may reduce their risk of coronary artery disease but at the same time run a higher risk of some other heart ailment.

Most doctors do not actively encourage moderate drinking, while keeping in mind that the practice may indeed be mildly beneficial. One reason is failure to agree upon what is moderate; another is the risk of drinkers becoming alcoholic. Also, Americans have not sloughed off puritanism entirely, and for many it seems somehow "wrong" to recommend that people drink "the devil's brew."

Alcohol relieves tension

Tension is supposed to be bad. Therefore, the reasoning goes, alcohol may be good. (Too much alcohol, needless to say, leads to accidents, marital spats, and tends to make one obnoxious, all sources of social tension.)

Alcohol may help diabetics

Diabetics are sometimes told not to drink, but, in fact, there is some evidence that dry, nonsweet wines and diluted distilled spirits may improve glucose tolerance and therefore be helpful in the treatment of diabetes. Also, diabetics tend to have low levels of HDLs, and alcohol, by correcting this defect, again may be useful. It should be pointed out that there is much controversy among internists on this point.

WHAT IS "NORMAL"?

When this chapter refers to effects of drinking, it generally refers to effects of moderate drinking (also called social or normal drinking). What *is* moderate drinking?

The definitions vary widely. The federal government recently studied the drinking habits of more than 30,000 households, dividing drinkers into abstainers, moderate drinkers, and heavy drinkers. A moderate drinker was defined as one who drank from four to thirteen drinks per week. A heavy drinker averaged two or more drinks per day, or over fourteen drinks per week. A drink was defined as a one-and-one-half-ounce highball, one bottle of beer, or four ounces of wine.

Using these definitions, the study found that 13 percent of men and 3 percent of women were heavy drinkers. Heavy drinkers were concentrated in the highest income groups (families earning $50,000 and over) and among whites. There is some inconsistency about the social-class correlation, with other studies reporting higher rates among blue-collar workers and urban blacks.

The study also found that abstainers had higher rates of certain diseases, particularly hypertension, heart disease, stroke, and diabetes. This seems to support the idea that drinking is good for your health. The truth, however, may be that healthy people tend to drink moderately, not that drinking moderately makes them healthy.

MYTHS AND MISCONCEPTIONS

Some other physical effects of alcohol should also be mentioned, if only because there are misconceptions about them.

Alcohol and urination

It is generally known that alcohol increases urination. It is generally not known that the increase is temporary, and that after a fairly short period of drinking the need to urinate decreases. On the morning after a night of heavy drinking, a person may not urinate at all. No explanation is available.

Dehydration

It is commonly believed that alcohol causes dehydration. It does not. When a person has a dry mouth and thirst after an evening of drinking, it may be because of the astringent effect of alcohol on the mucous membranes of the mouth. If anything, heavy drinkers may be slightly overhydrated because of the large volume of fluid they consume.

Alcohol and body temperature

It is generally known that alcohol produces a feeling of bodily warmth and therefore is just the thing for Saint Bernards to carry around their necks in caskets and for old boys to have at a frosty football game. Alcohol produces a feeling of warmth because it dilates blood vessels in the skin, which is why drinkers have red noses. However, the warmth can be harmfully illusionary. A person's resistance to the effects of severe cold, such as frostbite, is in no way increased by alcohol, although the victim may temporarily think it is.

So much for alcohol. The rest of the book concerns an effect of alcohol that remains truly mysterious, namely, why some people drink so much that they harm themselves and others. Let's next examine the condition called alcoholism.

CHAPTER TWO

Alcoholism

> In my judgment such of us who have never fallen victims [to alcoholism] have been spared more by the absence of appetite than from any mental or moral superiority over those who have. Indeed, I believe if we take habitual drunkards as a class, their heads and their hearts will bear an advantageous comparison with those of any other class.
>
> —Abraham Lincoln

> He drank, not as an epicure, but barbarously, with a speed and dispatch altogether American, as if he were performing a homicidal function, as if he had to kill something inside himself, a worm that would not die.
>
> —Baudelaire, writing about Edgar Allan Poe

In *The Lost Weekend,* Charles Jackson said that the alcoholic was a person who could take alcohol or leave it, so he took it. The National Council on Alcoholism defines the alcoholic as a person who cannot consistently predict how much he or she will drink, or for how long, once drinking begins. The distinguished alcoholism expert, E. M. Jellinek, said there were various types of alcoholics, including an "American-style" alcoholic who can refrain from drinking but cannot control the amount or duration of drinking once drinking begins, and a "French-style" alcoholic who can control the amount consumed but cannot abstain. (The two types are described in their own words later in the chapter.)

The currently "official" definition of alcoholism has been supplied by the American Psychiatric Association in the revised edition of *Diagnostic and Statistical Manual of Mental Disorders (DSM–III),* published in 1987, which lists the following nine characteristics of the alcoholic:

1. Alcohol often taken in larger amounts or over a longer period than the person intended.
2. Persistent desire or one or more unsuccessful efforts to cut down or control alcohol use.
3. A great deal of time spent in activities necessary to get alcohol, taking the substance, or recovering from its effects.
4. Frequent intoxication or withdrawal symptoms when expected to fulfill major role obligations at work, school, or home (e.g., does not go to work because hung over, goes to school or work intoxicated, or is intoxicated while taking care of his or her children), or when alcohol use is physically hazardous (e.g., drives when intoxicated).
5. Important social, occupational, or recreational activities given up or reduced because of alcohol.
6. Continued alcohol use despite knowledge of having a persistent or recurrent social, psychological, or physical problem that is caused or exacerbated by the use of the substance.
7. Marked tolerance: need for markedly increased amounts of the substance (i.e., at least a 50 percent increase) in order to achieve intoxication or desired effect, or markedly diminished effect with continued use of the same amount.
8. Characteristic withdrawal symptoms.
9. Alcohol often taken to relieve or avoid withdrawal symptoms.

According to *DSM–III,* any person who has three of the above manifestations can be considered "alcohol dependent." (*DSM–III* rejects the term *alcoholism* on the grounds that it contributes to the stigmatization of the condition.)

DSM–III also accepts the view that alcohol dependency may vary in severity. It defines degrees of severity as follows:

- *Mild:* Few, if any, symptoms in excess of those required to make the diagnosis, and the symptoms result in no more than mild impairment in occupational functioning or in usual social activities or relationships with others.
- *Moderate:* Symptoms or functional impairment between "mild" and "severe."
- *Severe:* Many symptoms in excess of those required to make the diagnosis, and the symptoms markedly interfere with occupational functioning or with usual social activities or relationships with others.
- *In Partial Remission:* During the past six months, some use of alcohol and some symptoms of dependence.
- *In Full Remission:* During the past six months, either no use of alcohol or use of alcohol and no symptoms of dependence.

These criteria were produced by a committee (to which the author belonged). Like all research criteria, they tend to be arbitrary. For the real flavor of alcoholism, real-life stories are the best source. Anyone who attends Alcoholics Anonymous (AA) meetings can hear these stories in rich abundance. Here is the kind of story one might hear:

> I am David. I am an alcoholic. I have always been an alcoholic. I will always be an alcoholic. I cannot touch alcohol. It will destroy me. It is like an allergy—not a real allergy—but *like* an allergy.
>
> I had my first drink at sixteen. I got drunk. For several years I drank every week or so with the boys. I didn't always get drunk, but I know now that alcohol affected me differently than other people. I looked forward to the times I knew I could drink. I drank for the glow, the feeling of confidence it gave me. But maybe that's why my friends drank too. They didn't become alcoholics. Alcohol seemed to satisfy some specific need I had, which I can't describe. True, it made me feel good, helped me forget my troubles, but that wasn't it. What was it? I don't know, but I know I liked it, and after a time, I more than liked it, I needed it. Of course, I didn't realize it. It was maybe ten or fifteen years before I realized it, *let* myself realize it.

My need was easy to hide from myself and others (maybe I'm kidding myself about the others). I only associated with people who drank. I married a woman who drank. There were always reasons to drink. I was low, tense, tired, mad, happy. I probably drank as often because I was happy as for any other reason. And occasions for drinking—when drinking was appropriate, expected—were endless. Football games, fishing trips, parties, holidays, birthdays, Christmas, or merely Saturday night. Drinking became interwoven with everything pleasurable—food, sex, social life. When I stopped drinking, these things, for a time, lost all interest for me, they were so tied to drinking. I don't think I will ever enjoy them as much as I did when drinking. But if I had kept drinking, I wouldn't be here to enjoy them. I would be dead.

So, drinking came to dominate my life. By the time I was twenty-five I was drinking every day, usually before dinner, but sometimes after dinner (if there was a "reason"), and more on weekends, starting in the afternoon. By thirty, I drank all weekend, starting with a beer or Bloody Mary in the morning, and drinking off and on, throughout the day, beer or wine or vodka, indiscriminately. The goal, always, was to maintain a glow, not enough, I hoped, that people would notice, but a glow. When five o'clock came, I thought, well, now it's cocktail hour and I would have my two or three scotches or martinis before dinner as I did on nonweekend nights. After dinner I might nap, but just as often felt a kind of wakeful calm and power and happiness that I've never experienced any other time. These were the dangerous moments. I called friends, boring them with drunken talk; arranged parties; decided impulsively to drive to a bar. In one year, at the age of thirty-three, I had three accidents, all on Saturday night, and was charged with drunken driving once (I kept my license, but barely). My friends became fewer, reduced to other heavy drinkers and barflies. I fought with my wife, blaming her for *her* drinking, and once or twice hit her (or so she said—like many things I did while drinking, there was no memory afterward).

And by now I was drinking at noontime, with the lunch hour stretching longer and longer. I began taking off whole afternoons, going home potted. I missed mornings at work because of drinking the night before, particularly Monday mornings. And I began drinking weekday mornings to get going. Vodka and orange juice. I thought vodka wouldn't smell (it did). It usually lasted until an early martini luncheon, and I then suffered through until cocktail hour, which came earlier and earlier.

By now I was hooked and knew it, but desperately did not want others to know it. I had been sneaking drinks for years—slipping out to the kitchen during parties and such—but now I began hiding alcohol, in my desk, bedroom, car glove compartment, so it would never be far away, ever. I grew panicky even thinking I might not have alcohol when I needed it, which was just about always.

For years, I drank and had very little hangover, but now the hangovers were gruesome. I felt physically bad—headachy, nauseous, weak—but the mental part was the hardest. I loathed myself. I was waking early and thinking what a mess I was, how I had hurt so many others and myself. The words "guilty" and "depression" sound superficial in trying to describe how I felt. The loathing was almost physical—a dead weight that could be lifted in only one way, and that was by having a drink, so I drank, morning after morning. After two or three, my hands were steady, I could hold some breakfast down, and the guilt was gone, or almost.

Despite everything, others knew. There was the odor, the rheumy eyes and flushed face. There was missing work and not working well when there. Fights with wife, increasingly physical. She kept threatening to leave and finally did. My boss gave me a leave of absence after an embarrassed remark about my "personal problems." At some point I was without wife, home, or job. I had nothing to do but drink. The drinking was now steady, days on end. I lost appetite and missed meals (besides, money was short). I awoke at night, sweating and shaking, and had a drink. I awoke in the morning vomiting and had a drink. It couldn't last. My ex-wife found me in my apartment shaking and seeing things, and got me in the hospital. I dried out, left, and went back to drinking. I was hospitalized again, and this time stayed dry for six months. I was nervous and couldn't sleep, but got some of my confidence back and found a part-time job. Then my ex-boss offered my job back and I celebrated by having a drink. The next night I had two drinks. In a month I was drinking as much as ever and again unemployed. That was three years ago. I've had two big drunks since then but don't drink other times. I think about alcohol and miss it. Life is gray and monotonous. The joy and gaiety is gone. But drinking will kill me. I know this and have stopped—for now.

To summarize:

A tree is known by its fruit; alcoholism by its problems. An alcoholic is a person who drinks, has problems from drinking, but goes on drinking anyway. This may be the best definition of alcoholism that exists. Theoretically, a person can drink a gallon of whiskey a day for a lifetime, not have problems, and therefore not be alcoholic. Theoretically. In fact, heavy drinkers almost always have problems. Sometimes they are mild. Alcohol calories may result in overweight—a cosmetic if not a medical problem. A heavy drinker may say things while drinking that would not or should not be said other times. A minor traffic offense may have major consequences when the driver's breath has alcohol on it.

Problems, yes, but alcoholism? The verdict rests with the observer. A fundamentalist teetotaler may view any problem from drinking as alcoholism. Moderate drinkers may be more indulgent, saying in effect, "These things happen. If they do not happen too often, it probably does not mean much." But what is too often? Except in extreme cases—the Davids, about whom everyone agrees—there will always be controversy about who is and who is not an alcoholic. This is understandable; doctors disagree about who has heart disease if the case is mild.

"Alcoholism" in this book refers to the David type of alcoholism, granting that patterns of human behavior are bewilderingly variable, even patterns of illness. Not all Davids, for example, reach bottom (in AA terms). Some stop drinking long before. Others drink, but with enough control to prevent the big problems from happening. The essence of the David type of alcoholism is a vulnerability to alcohol that sets him apart from other drinkers. By taking extreme measures, such as total abstinence, he may prevent alcohol problems; but if he drinks at all, the chance of developing problems is high, and this vulnerability appears to be lifelong.

How many people have this condition? We don't know. Population surveys show that about 70 percent of adults drink. About 9 percent have problems from drinking, mostly minor; another 9 percent have had problems in the past. (There seems to be a considerable migration in and out of the "problem-drinking" pool.) Among the problem drinkers are a subgroup called alcoholics, as defined above. A few years ago there were said to be five million American alcoholics. A more recent estimate is fourteen million. This increase coincides with increased efforts by the federal government to study and treat alcoholism. The connection is probably coincidental, but since the figures are fictitious anyway, it does not matter. No one doubts that alcoholics like David exist in large numbers.

Alcohol problems fall into three groups: psychological, medical, and social.

PSYCHOLOGICAL PROBLEMS

Preoccupation with alcohol

The alcoholic thinks about alcohol from morning till night, and at night, if not too drunk to dream, dreams about alcohol. When to have the first drink? When the next? Remember bars are closed on election day. Remember liquor stores are closed on Sundays. Prepare, prepare. Will they sell more than two drinks on the airplane? Take a flask. Do the Smiths drink? Find out before accepting their dinner invitation. This goes on and on, blotting out other thoughts, other plans.

It is obsessional in precisely the way psychiatrists use the word. Obsessions breed compulsions, and when an alcoholic drops into a bar or liquor store, ever so casually, it is as compulsive as the neurotic washing his hands for the twentieth time that day.

Self-deception

But he must not admit it to himself. "We are all victims of systematic self-deception," Santayana said, and the alcoholic is a victim par excellence. People are victims of many things—cancer, lust, society—and can accept it. But, deep down, the alcoholic believes he is doing it to himself; he is the perpetrator, not the victim. And this he cannot accept, so he lies to himself.

"I can stop drinking anytime. Important people drink. Churchill drank. Today is special—a friend is in town. Nothing is going on—why not? Life is tragic—why not? Tomorrow we die—why not?"

As he lies to himself, he lies to others, and concealment becomes a game like the one children play when they raid the cookie jar and hope their mother won't notice.

Guilt

But he does know and can't help knowing. There are too many reminders. The wife's pleas and tantrums. The boss's "friendly" advice. The crumpled fender. The night terrors and night sweats. The trembling hands. The puffy eyes and blotchy complexion. The terrifying memory gaps. All spell self-destruction, and even the cleverest self-deceiver knows it.

Amnesia

Alcoholics have memory lapses when they drink, and this is often attributed to guilt. It is said the forgetter does not want to remember. Nonalcoholics also have memory lapses when they drink, not so often or so severely, but nonalcoholics by definition drink less. Memory lapses—or blackouts, as they are called when alcohol is involved—are probably not due to guilt. More likely alcohol, in some people on some occasions, interferes with chemical processes that make memory—perhaps the most mysterious of biological phenomena—possible.

Precisely how it occurs is unknown, but the memory lapses

are genuine. The drinker does things when he is drinking that ordinarily he would remember perfectly, but when he sobers up, usually the next day, he has no recollection of what he has done. Sometimes he realizes that he had a memory lapse. He is apprehensive. He checks to see if his car is in the garage. He looks for dents that weren't there before. His overriding fear is that he did something—broke a law, harmed someone—and punishment is at hand. He retraces his movements of the night before. "Was I here, Joe?" he asks. Told that he was: "What did I do? Was I drunk?" Reassured that he did nothing wrong and was no more drunk than usual, he goes to the next place where he might have been. Alternately, he may avoid all places and all companions he might have visited or been with during the forgotten interval, preferring not to know.

In truth, people rarely do things during blackouts that they don't also do when they are drunk and suffer no memory loss.

During blackouts, the person is conscious and alert. He may appear normal. He may do complicated things—converse intelligently, seduce women, travel. A true story:

> A thirty-nine-year-old salesman awoke in a strange hotel room. He had a mild hangover but otherwise felt normal. His clothes were hanging in the closet; he was clean-shaven. He dressed and went down to the lobby. He learned from the clerk that he was in Las Vegas and that he had checked in two days previously. It had been obvious that he had been drinking, the clerk said, but he hadn't seemed very drunk. The date was Saturday the 14th. His last recollection was of sitting in a St. Louis bar on Monday the 9th. He had been drinking all day and was drunk, but could remember everything perfectly until about 3 P.M., when "like a curtain dropping," his memory went blank. It remained blank for approximately five days. Three years later, it was still blank. He was so frightened by the experience that he abstained from alcohol for two years.

Some people forget and do not realize when sober that they have forgotten anything. Someone tells them and then they remember a little.

> A fifty-three-year-old member of Alcoholics Anonymous said that he had experienced many blackouts during his twenty-five years of heavy drinking. He could not remember his first blackout, but guessed it had happened about fifteen years before. The memory loss had not bothered him, he said; he assumed everyone who drank had trouble with memory. Sometimes, however, it was embarrassing to be told he had said something or gone somewhere and not recalled it. Upon being told, he would sometimes remember the event and sometimes not. Occasionally, months later, something would remind him of the event and his memory would "snap back." Typically, he could remember some parts of a drinking episode and not others; a half hour might be blanked out and the next hour remembered. The forgotten parts appeared to have no more emotional significance than the remembered ones. "It's like turning a switch on and off."

Sometimes a curious thing happens when a person is drinking: The drinker recalls things that happened during a previous drinking period which, when sober, he had forgotten. For example, alcoholics often report hiding money or alcohol when drinking, forgetting it when sober, and having their memory return when drinking again. This is reminiscent of the state-dependent learning described in the first chapter. Whatever the explanation, the mind does play odd tricks on drinkers:

> A forty-seven-year-old housewife often wrote letters when she was drinking. Sometimes she would jot down notes for a letter and start writing it but not finish it. The next day, sober, she would be unable to decipher the notes. Then she would start drinking again, and after a few drinks the meaning of the notes would become clear and she would resume writing the letter. "It was like picking up the pencil where I had left off."

Anxiety and depression

What goes up comes down, and alcoholic euphoria is followed by alcoholic depression with a kind of Newtonian inevitability. Anxiety and depression occur not only with hangovers but intermittently during the drinking period itself, if the drinking is heavy and continuous. This sequence is common: a man feels bad for any reason (it's a gray day); he drinks, feels better; then he feels bad again, this time because the alcohol effect is wearing off; he drinks again, feels better again. And a vicious cycle is under way, based on alcohol's ability to alternately raise and lower spirits. This rollercoaster effect is probably chemical in nature, but the drinker only knows that alcohol, having raised his spirits, now lowers them, and that the best way to raise them again is to have another drink.

MEDICAL PROBLEMS

If this process goes on long enough, there are usually medical problems. They may take years to develop, and some lucky drinkers never are affected. Incredibly, a person may consume a ferocious quantity of alcohol, maybe a fifth or a quart of whiskey a day for twenty years or longer, and when he dies a "natural" death, his brain, liver, pancreas, and coronary arteries appear normal. But the odds are strong that something will give. Here are some favorite targets:

The stomach

Gastritis, inflammation of the stomach's lining, is common. The symptoms are gas, bloating, heartburn, nausea. The cure is alcohol. Before-and-after pictures of the stomach prove it. Raw and inflamed from a night, or week, of heavy drinking, the stomach is miraculously restored to a normal appearance after a shot or two of alcohol.

Alcohol's role in producing stomach or duodenal ulcers is debatable. Ulcers are caused by hydrochloric acid and digestive enzymes. These are powerful enough to digest fish bones and the toughest beefsteak, but inexplicably do not digest the stomach itself. The mucous lining of the stomach somehow prevents it. When the protection is lost, ulcers develop. If alcohol, hot peppers, or pent-up anger produce ulcers at all, they must do so either by increasing acid production or by breaching the protective barrier. But alcohol, in strong doses, decreases rather than increases acid production. Its effect on the protective lining is not clear.

The liver

The word *cirrhosis* comes from the Greek word for yellow-orange, probably because people with cirrhosis become jaundiced. Alcoholics are disposed to a type of cirrhosis called portal cirrhosis, or Laennec's cirrhosis.

In the first stages of cirrhosis, liver cells become inflamed and gradually die out. The liver swells up and can be felt through the belly wall, whereas usually it is hidden behind the ribs. New cells appear—the body incorrigibly bent on keeping things going—but with a difference. Previously the cells lined up in columns, forming banks for concentric canals through which blood coursed. The new cells form higgledy-piggledy, and the blood flow comes nearly to a halt.

The results are predictable. Blood backs up and seeps into the abdomen, which swells like a balloon, or it detours around the liver, engorging the paper-thin veins of the esophagus. If the veins burst, fatal hemmorhage may result. The liver cells, formerly little factories with many functions, go on strike and their production of proteins, blood-clotting factors, and other vital constituents falls off. In men, the cells no longer suppress female sex hormones (the manliest man has some female sex hormones), so men's breasts grow, their testicles shrink, and they lose their baritone voices, beards, and hairy chests.

As the process continues, scar tissue forms and eventually the liver is like a small lumpy rock, incapable of sustaining life.

Cirrhosis is the ninth leading cause of death in the United States. Until recently it was believed that most people with Laennec's cirrhosis were alcoholics, but a recent study suggests otherwise. Some cirrhosis patients in the study drank no alcohol at all. A third consumed less than two drinks per day.

Studies are inconsistent on this point, but one fact is uncontested: Most alcoholics do not develop cirrhosis. The connection between drinking and cirrhosis is still not understood.

Nerve fibers

The long nerve fibers extending from the spinal cord to muscles often suffer degenerative changes in alcoholics. The fibers make muscles contract and maintain muscular tone; they also transmit back to the spinal cord, and thence the brain, messages from sensory receptors in muscle and skin. The degeneration of nerves results in muscular weakness and eventual wasting and paralysis. Pain and tingling are experienced, and there may be eventual loss of sensation. The cause of nerve-fiber degeneration in the alcoholic is not alcohol. It is a vitamin deficiency. High doses of B vitamins almost always restore the fibers to their normal state, if not given too late.

Brain damage

Whether excessive use of alcohol produces atrophy of the brain surface (cortex) has been debated for many years. Computer tomography (CT) scans of alcoholics have been contradictory, most showing some atrophy, a few showing none, and one showing reversible atrophy. Some experts hold that what appears to be atrophy may instead represent alterations caused by temporary fluid shifts, causing the ventricles to swell and the cortex to appear thinner. This view is consistent with evidence that the body becomes water-logged from excessive drinking. The old term "wet brain"—a largely disused term previously applied to the brains of alcoholics—may describe an actual condition.

Other than the CT studies, there is no direct evidence that alcohol alone causes brain damage. After many years of heavy drinking, most alcoholics, when recovering from their latest drinking bout, show little or no sign of intellectual impairment. Their IQs are normal, their thoughts logical, and their minds clear. If there is an impairment, it is usually subtle, rarely persists, and can be attributed to factors other than loss of brain cells—poor motivation, for example, in taking the tests psychologists are forever giving alcoholics.

A small minority of alcoholics do suffer brain damage, but apparently the cause is a deficiency of thiamine, a B vitamin, rather than a toxic effect of alcohol itself.

The malnourished alcoholic gets too little thiamine, and if the deprivation persists and is severe, certain well-demarcated areas of the brain may be destroyed. These areas are, among other things, involved in memory storage. Their destruction results in severe memory impairment. A German named Wernicke and a Russian named Korsakoff first described the disease. The patient with Wernicke–Korsakoff disease—almost always an alcoholic in this country—can remember the distant past fairly well, has a normal IQ, and seems reasonably bright except for one problem: He cannot remember anything that happened to him a few minutes after it happens. The condition is devastating, and the chronic Wernicke–Korsakoff patient needs custodial care for the rest of his life. Thiamine, given early, may prevent a permanent defect. Fortunately, the condition is rare.

Alcoholics also are inclined to suffer degenerative changes in the cerebellum, the half-melon bulge at the base of the brain that regulates coordination. An unsteady gait results. Vitamin deficiency is believed to be the cause.

Impotency

> MACDUFF: What three things does drink especially provoke?
> PORTER: Merry, sir, nose-painting, sleep and urine. Lechery, sir, it provokes, and unprovokes; it provokes the desire, but it takes away the performance. . . .
>
> *Macbeth* (act 2, scene 3)

Shakespeare names not three but four of alcohol's well-known actions, lechery being the most famous. By dilating blood vessels, alcohol "paints" the nose; it makes people sleepy; and, one of the things a novice drinker first notices about drink, it makes him go to the bathroom.

It also increases sexual desire. More accurately, perhaps, it "releases" sexual desire—the well-known disinhibiting effect of alcohol. But performance may be impaired. Drunken men have trouble achieving an erection or ejaculating. Whether sexual performance in drunken women also is impaired is hard to determine. Women have less to erect, and the female orgasm remains poorly understood.

Some alcoholics not only have trouble with sexual performance when drinking; the problem persists long into sobriety. Whether the cause is psychological or physical is not known, but it may contribute to a well-described syndrome: alcoholic conjugal paranoia. Husbands, without evidence, become convinced that their wives are unfaithful. They hound their wives, accuse them, search for anything to support their delusion: inspect underwear for semen spots, hire detectives, sniff blouses for aftershave lotion. The delusion is precisely that: a fixed false idea. It may be related to the husband's feelings of inadequacy about his own sexual ability and perhaps to feelings of inadequacy about life in general, in ruins from years of heavy drinking. Whether women alcoholics also develop alcoholic paranoia is not clear.

Other medical problems

"To know syphilis is to know medicine," William Osler said at a time when syphilis was untreatable and affected, or could affect, nearly every organ in the body. The same can be said of alcoholism. It can affect, or is alleged to affect, nearly every organ. Rare is the medical journal that does not occasionally publish new evidence of alcohol's dangers: heart disease, muscle disease, pancreatitis, anemia, cancer, not to mention the conditions described above. The list is long and growing. The problem is: Are the reports reliable and is the culprit alcohol?

Take the case of cancer. Heavy drinkers apparently have more cancer of the mouth and throat than do nonheavy drinkers. But heavy drinkers are almost always heavy smokers, and heavy smokers also have a higher than average incidence of cancer of the mouth and throat. Alcohol alone will not cause cancer, but in some poorly understood way it may enhance the carcinogenic effects of tobacco smoke.

Recent evidence implicates alcohol in three serious illnesses heretofore overlooked: hypertension, stroke, and breast cancer. A word about each.

Studies on hypertension are contradictory, but the following conclusion tentatively can be drawn: Excessive drinking causes a temporary elevation in blood pressure, with a return to normal levels within a week after the person stops drinking. Where hypertension persists, the cause is probably not alcohol. In one study, formerly heavy drinkers had blood pressure readings comparable with those of teetotalers. In the face of conflicting data, the word *tentative* must be strongly emphasized.

Heavy drinkers are reported to have an increased incidence of strokes. Alcohol may or may not be directly responsible. It has been suggested that perhaps alcoholics who have high blood pressure (independent of their drinking) may be at increased risk for stroke because they do not cooperate with doctors in taking medications to control blood pressure. The nature of the relationship remains to be determined.

Two articles published in the May 1987 issue of the *New England Journal of Medicine* reported that women who used alcohol in *any* amount had a 50 percent higher chance of developing breast cancer than those who did not drink alcohol at all. An association is not necessarily a cause. In one of the largest studies, the breast cancer rate was 71 per 10,000 for those who used alcohol in any amount. It was 59 per 10,000 for those who used no alcohol. Clearly, therefore, use of alcohol cannot be considered a primary cause of breast cancer.

Since drinkers are more often smokers than nondrinkers, it has been suggested that perhaps smoking was the actual cause of the increase in breast cancer. However, one study suggests that women who smoke cigarettes have a *lower* rate of breast cancer than do nonsmokers. In any case, based on these studies, the association between drinking and breast cancer applies to light as well as heavy drinkers and remains a subject of concern.

The problem in ascribing an illness to heavy drinking is that heavy drinkers differ from nonheavy drinkers in other ways. As mentioned, they smoke more. They often eat less. They often lead irregular lives—staying up all hours, never exercising, sleeping it off in doorways. How can these potentially harmful influences be separated from the effects of alcohol? It is difficult.

WITHDRAWAL

Alcoholics also experience a medical problem that, strictly speaking, does not come from drinking alcohol but from *not* drinking alcohol. This is the alcohol-withdrawal syndrome. It is commonly, but mistakenly, called the DTs, or delirium tremens. In medical usage "delirium" means gross memory disturbance, usually combined with insomnia, agitation, hallucinations, and illusions. Most alcoholics do not experience delirium.

As a rule, alcohol withdrawal is a distressing but temporary condition lasting from two days to a week. The mildest symptom is shakiness, which begins a few hours after the patient stops drinking, sometimes awakening him during sleep. Morning shakes are inevitable if the drinker has been drinking enough. His eyelids flutter, his tongue quivers, but, most conspicuously, his hands shake, so that transporting a cup of coffee from saucer to mouth is a major undertaking.

The cure for the shakes, as for all alcohol-withdrawal symptoms, is a drink or two.

After a day or two without drinking, the alcoholic coming off a bender may start hallucinating—seeing and hearing things that others do not see or hear. He often realizes he is hallucinating and blames alcohol. Not always, however. Sometimes the hallucinations are vivid, frightening, and as real as life.

Occasionally alcoholics have convulsions that resemble the

grand-mal seizures of the epileptic. Most alcoholics are not epileptic and have seizures only when withdrawing from alcohol. The seizures usually occur one to three days after the person stops drinking.

The most severe form of withdrawal involves delirium and justifies using the term delirium tremens, the "tremens" referring to the shakiness. Delirium is ominous. It often means the person has not only withdrawal symptoms but also a serious medical illness, often of the type to which alcoholics, because of their way of living, are vulnerable: pneumonia, fractures, blood clots in the brain, liver failure. People occasionally die in delirium tremens, whereas death from milder forms of withdrawal is rare.

Many alcoholics are capable of withdrawing from alcohol on their own. They often do this by tapering off—gradually decreasing the amount they drink. Serious withdrawal symptoms, however, justify hospitalization so that tranquilizers can be given to make the alcoholic feel better, vitamins to prevent brain damage, and frequent medical examinations to exclude medical illness.

The DTs have been described brilliantly in fiction by, among others, Malcolm Lowry and Mark Twain. Lowry, in his novella "Lunar Caustic," wrote from personal experience how it felt to wake up in an alcoholic ward:

> The man awoke certain that he was on a ship. If not, where did those isolated clangings come from, those sounds of iron on iron? He recognized the crunch of water pouring over the scuttle, the heavy tramp of feet on the deck above, the steady Frère *Jac*ques: Frère *Jac*ques of the engines. He was on a ship, taking him back to England, which he never should have left in the first place. Now he was conscious of his racked, trembling, malodorous body. Daylight sent probes of agony against his eyelids. Opening them, he saw three negro sailors vigorously washing down the deck. He shut his eyes again. Impossible, he thought . . .
>
> As day grew, the noise became more ghastly: what sounded like a railway seemed to be running just over the ceiling. An-

other night came. The noise grew worse and, stranger yet, the crew kept multiplying. More and more men, bruised, wounded, and always drunk, were hurled down the alley by petty officers to lie face downward, screaming, or suddenly asleep on their hard bunks.

He was awake. What had he done last night? Nothing at all, perhaps, yet remorse tore at his vitals. He needed a drink desperately. He did not know whether his eyes were closed or open. Horrid shapes plunged out of the blankness, gibbering, rubbing their bristles against his face, but he couldn't move. Something had got under his bed too, a bear that kept trying to get up. Voices, a prosopopoeia of voices, murmured in his ears, ebbed away, murmured again, cackled, shrieked, cajoled; voices pleading with him to stop drinking, to die and be damned. Thronged, dreadful shadows came close, were snatched away. A cataract of water was pouring through the wall, filling the room. A red hand gesticulated, prodded him: over a ravaged mountain side a swift stream was carrying with it legless bodies yelling out of great eye-sockets, in which were broken teeth. Music mounted to a screech, subsided. On a tumbled bloodstained bed in a house whose face was blasted away a large scorpion was gravely raping a one-armed negress. His wife appeared, tears streaming down her face, pitying, only to be instantly transformed into Richard III, who sprang forward to smother him.

After a few days, the DTs go away. Lowry's patient "now knew himself to be in a kind of hospital, and with this realization everything became coherent and fell into place. The sound of water pouring over the scuttle was the terrific shock of the flushing toilets; the banging of iron and the dispersed noises, the rattling of keys, explained themselves; the frantic ringing of bells was for doctors or nurses; and all the shouting, shuffling, creaking and ordering was no more than the complex routine of the institution."

Psychiatric patients are rarely dangerous, but delirious patients are an exception. They may be dangerous indeed, as was the case of Huckleberry Finn's alcoholic father, whose DTs were described by Mark Twain as follows:

I don't know how long I was asleep, but all of a sudden there was an awful scream and I was up. There was Pap looking wild, and skipping around and yelling about snakes. I couldn't see no snakes, but he said they was crawling up his legs; and then he would give a jump and scream, and say one had bit him on the cheek. I never see a man look so wild. Pretty soon he was all fagged out, and fell down panting; then he rolled over and over, screaming and saying there was devils a-hold of him. He wore out by and by, and laid still awhile, moaning. Then he laid stiller, and didn't make a sound. I could hear the owls and the wolves away off in the woods, and it seemed terrible still. He was laying over by the corner. By and by he raised up partway and listened, with his head to one side. He wails, very low:

"Tramp-tramp-tramp; that's the dead; tramp-tramp-tramp; they're coming after me; but I won't go. Oh, they're here! Don't touch me—don't. Hands off—they're cold; let go. Oh, let a poor devil alone!"

He rolled himself up in his blanket and went to crying. But by and by he rolled out and jumped up to his feet looking wild, and he see me and went for me. He chased me round and round the place with a clasp knife, calling me the Angel of Death, and saying he would kill me. . . . I begged, and told him I was only Huck; but he laughed such a screech laugh, and roared and cussed, and kept on chasing me. Once when I turned short and dodged under his arm he got me by the jacket between my shoulders, and I thought I was gone; but I slid out of the jacket and saved myself. Pretty soon he was all tired out, and dropped down with his back against the door, and said he would rest a minute and then kill me. He put his knife under him, and pretty soon he dozed off.

SOCIAL PROBLEMS

In simpler times, it was said that marijuana smoking was a "crime without a victim," but even then no one would have called alcoholism a victimless "crime." The victims of alcoholism are legion: spouses, children, other relatives, bosses, fellow workers, pedestrians, drivers, police, judges, physicians who get called late at night, taxpayers who often pick up the bill for treatment, and other innocent and not so innocent people who cross the alcoholic's path.

"There are lies, damn lies, and statistics," Mark Twain said, but in the case of alcoholism the statistics are probably more true than not. They include the following:

1. The average city police officer spends half of the time dealing with alcohol-related offenses. Nearly half of the men and women in prisons are alcoholic or, at any rate, heavy drinkers. Most murderers are drinking at the time they commit a murder, and so are most of the victims, although how many would be considered alcoholic is uncertain.
2. Between 20 and 30 percent of male psychiatric admissions are alcoholic or have alcohol-related problems. About one quarter of the men admitted to general hospital wards for medical treatment have alcohol-related problems.
3. Industry loses at least several billion dollars a year because of absenteeism and work inefficiency related to alcoholism. Monday morning and Friday afternoon absenteeism, at least partly attributable to alcoholism, is so common that both industry and unions are considering a four-day work week (whereupon Tuesday morning or Thursday afternoon absenteeism will probably become common).
4. Alcoholics are about three times more likely to be divorced than nonalcoholics. The common explanation is that alcoholics drink too much, the spouse complains about it, and separation and divorce are the outcome. There may be another explanation, discussed in chapter 4.

   Alcoholics have a death rate three or four times higher than nonalcoholics. The most common causes, aside from medical diseases, are accidents and suicides. There are an estimated 15,000 deaths a year from alcohol-related automobile accidents. Studies indicate that most of the drinking drivers are not just social drinkers coming home from a Christmas party but serious problem drinkers, alcoholic by most definitions. About one out of four suicides in the United States is an alcoholic, usually a man over thirty-five.

NATURAL HISTORY

Alcoholism has a natural history. This means the condition tends to develop at certain ages, progresses in a more or less predictable manner, and terminates in more or less predictable ways. "More or less" is an important qualifier, as there is much variation. Men and women vary; whites and blacks; Americans and French.

The "typical" white male American alcoholic begins drinking heavily in his late teens or early twenties, drinks more and more throughout his twenties, starts having serious problems in his thirties, is hospitalized for drinking (if ever) in his mid- or late thirties, and is clearly identified by himself and others as alcoholic—a man who cannot drink without trouble—between age forty and fifty.

Men, with rare exceptions, do not become alcoholic after forty-five. There is an "age of risk" for alcoholism, as for most illnesses, and if a man has no symptoms of alcoholism by his late forties, he probably will develop none.

The illness ends by death from suicide, accident, or medical illness—or by cessation of drinking. Few alcoholics return to social drinking. The American white male alcoholic is a person who can abstain from alcohol for long periods, perhaps most of a lifetime, but once drinking begins he runs a heavy risk of losing control of the amount he drinks. The loss of control may be gradual. His drinking may slowly escalate for weeks or months, but at some point alcohol comes to dominate his life.

Black alcoholics start drinking younger—often in their early and mid-teens. By twenty they may be floridly alcoholic and need hospitalization. They have withdrawal hallucinations more often than do white alcoholics and, for unknown reasons, are less suicidal in middle age.

George Vaillant, a professor at Dartmouth, published a series of papers and a book during the 1980s shedding much light on the "natural history" of alcoholism. Based on a forty-year follow-up study of 500 men originally studied as teenagers, Vaillant came up with some surprises.

It is often stated, for example, that alcoholics, from early childhood, live in a state of insecurity. Vaillant found that the pre-alcoholic personalities of most alcoholics differed little from nonalcoholics; if anything, those who became alcoholic were more assertive, more self-confident, and less anxious. Subjects with oral and dependency traits (thumb-sucking, compulsive eating) were actually less likely to develop alcoholism.

Treatment didn't seem to alter the course of the illness. "Stable abstinence," Vaillant wrote, "occurred most often in untreated and severely alcohol-dependent individuals. . . . Until an alcohol abuser becomes very symptomatic, the subjective pain is not sufficiently severe to lead to recovery." This sounds very much like the AA notion that alcoholics must "reach bottom" before they can recover.

It is commonly believed that alcoholics come from unstable families. Vaillant found this was true only if the instability was caused by an alcoholic parent. Backgrounds of unstable families without an alcoholic parent did not lead to alcoholism in Vaillant's subjects. Unstable families with alcoholism in the family were associated with alcoholism in the children—a relationship just as likely based on heredity as on instability. Vaillant concluded that a "difficult life was rarely a major reason why someone developed alcohol dependency."

Vaillant found little connection, in fact, between alcoholism and mental health: "The three childhood variables that most powerfully predicted positive adult mental health—boyhood competence, warmth of childhood, and freedom from childhood emotional problems—did not predict freedom from alcoholism; whereas the three variables that most powerfully predicted alcoholism—a family history of alcoholism, ethnicity, and adolescent behavior problems—did not predict poor mental health." By ethnicity he really meant the Irish subjects in his study, who had a disproportionately high rate of alcoholism.

Vaillant also challenged the common belief that alcoholism is an insidious, progressive disease that, if not treated, eventually ends in death. Many alcoholics, he found, recover with or without treatment. In fact, as many recover without treatment as with. (This finding distressed Vaillant, the therapist, but he consoled himself with the reflection that treatment probably did no harm.)

More will be said in chapter 3 about the connection, if any, between psychiatric illness and alcoholism. Meanwhile, it should be recognized that alcoholism comes in a variety of packages, strongly influenced by culture. For example, there is Jellinek's distinction between the American and French-type alcoholics alluded to earlier.

A French alcoholic describes himself:

> My name is Pierre. I am not an alcoholic. I do not know alcoholics. There are no alcoholics in France, except tourists.
>
> I have drunk wine since I was a child. Wine is good for you. I drink it with meals and when I am thirsty. Since I was a young man, I have drunk three or four liters of wine every day. I also enjoy an occasional aperitif, especially on Sunday mornings and after work. I never drink more than this. I have no problems from alcohol.
>
> Once, when I was in the army, no wine was permitted. I started shaking all over and thought bugs were crawling on me. I think it was the army food. My doctor says my liver is too large. My father and grandfather had large livers. It probably means nothing.

The American alcoholic stereotype has two choices—abstain or go on a bender. The French alcoholic stereotype does not go on benders, but cannot abstain.

WOMEN AND ALCOHOL

In the last few years some tantalizing distinctions have been made between male and female drinkers and alcoholics. The most talked about, if not most important, distinction arises from the woman's ability to have children. There is now evidence that drinking has adverse effects on pregnancy.

Here are ten ways in which women alcoholics appear to differ from men alcoholics:

1. Women tend to become alcoholic at an older age. If men are not alcoholic by their mid-forties, they probably will not become alcoholic. This is less true of women.
2. Women alcoholics are more likely to have a depressive illness preceding or coinciding with heavy drinking.
3. Alcoholism in women is more serious. Women are harder to treat and to stay sober for briefer intervals. The interval between the onset of heavy drinking and the start of treatment appears to be shorter for women. (Women are more likely than men to seek help for health problems in general, and to do so at an earlier stage, so this "telescoping" in the development of alcoholism in women may have no connection with the illness itself.)
4. Women alcoholics more often have disruptive early life experiences, such as loss of a parent or other close relative, or psychiatric problems in the family. In several studies about half of women alcoholics reported a parent missing during their childhood, compared to about 15 percent for male alcoholics.
5. Women alcoholics more often have alcoholism in the family. Many studies have shown a high incidence of alcoholism (upwards of 60 percent) in a parent or sibling of women alcoholics. Women alcoholics are more likely to have relatives who are clinically depressed or who commit suicide.
6. Women alcoholics tend to use alcohol medicinally as a form of self-treatment. Male alcoholics tend to use alcohol (at least early on) recreationally and socially. Women alcoholics are more likely to use prescribed as well as over-the-counter medications. They are more likely to leave a doctor's office with a tranquilizer prescription. They are more likely to seek out solutions in medications, and apparently doctors encourage this tendency.
7. Women are more likely to cite a traumatic event as the cause of heavy drinking. The event may be a divorce, rejection by a spouse or lover, abandonment, the death of some-

one close, hysterectomy, miscarriage, or other health problem.

8. Women alcoholics tend to be submissive as children, rebellious as adults. This is said to be less true of men alcoholics.
9. Men generally are introduced to heavy drinking by other men when they are young. Women tend to get involved in heavy drinking later in life, often through the influence of husbands or men friends.
10. Based on a study of members of Alcoholics Anonymous, women are more likely to have a personality change when drinking, find unexplained bruises after a drinking episode, or drink before a "new situation."

Some of these distinctions are pretty conjectural and probably some do not exist. But all have been reported in respectable scientific journals in the past ten years.

Items six and eight are particularly conjectural. Item six was based mainly on studies by men; the facts may be accurate, but the interpretation biased. Regarding item eight, if every submissive child who became rebellious as an adult drank too much, the oceans could not contain enough alcohol to meet the demand. One wonders, too, about the bruises mentioned in item ten. Maybe men are less aware of their bruises because of their hairy legs.

The scientific journals yield even more information about gender and alcoholism. For example, more men are alcoholic than women. Some believe the gap is closing, but there is no direct evidence for this. The best current estimate of a sex ratio for alcoholism is 3:1 (male:female).

The number of women drinkers is on the rise. In secondary schools, as many girls now drink as boys. This was not true several years ago. But has there been an increase in heavy drinking among women? Probably not. Less than 10 percent are heavy drinkers, meaning they drink almost every day and get intoxicated perhaps several times a month. Between 20 and 40 percent of men fall in this category, with young men more likely than older men to be heavy drinkers. (The exact figures vary with the study, but the trends remain the same.)

Because more girls are drinking than apparently ever before, it is commonly assumed that the rate of heavy drinking and alcoholism will automatically increase in women. This may not occur for one reason: More women than men are physiologically intolerant of alcohol; after a modest amount of alcohol, more women experience dizziness, headache, nausea, or a sense of simply having enough. Since this "protection" is physiological and probably genetically determined, presumably it will not disappear even though more secondary school girls drink now and then and society is more tolerant of drunken behavior in women.

Men and women have different hormones, and hormones in women, if not men, seem to influence drinking behavior. When drinking the same amount of alcohol, in proportion to body weight, women have higher blood levels of alcohol than men. These blood levels vary according to the phase of the hormonally controlled menstrual cycle, being highest in the premenstrual period. Many women say they drink more during the premenstrual period, and alcoholic women more often give a history of premenstrual tension than do nonalcoholic women. Just how these hormonal shifts relate to alcoholism, if they relate at all, is not known.

The fetal alcohol syndrome

Women have heard for a long time that they shouldn't drink during pregnancy. The idea goes back at least to biblical times. In Judges 13:7, an angel tells Samson's mother: "Behold, thou shalt conceive and bear a son: and now drink no wine or strong drink."

In early Carthage, bridal couples were forbidden to drink for fear of producing a defective child. According to Aristotle, "Foolish, drunken, and harebrained women most often bring forth children like unto themselves, morose and languid." In 1834 a report to the British House of Commons said: "Infants of alcoholic mothers often have a starved, shrivelled, and imperfect look."

The biblical injunction does not make clear why pregnant

women shouldn't drink. The concern may have been that the child would become alcoholic. Plutarch said that "Drunkards beget drunkards," reflecting a belief held well into the twentieth century that traits and habits acquired by parents would be passed along to offspring. There was also, of course, the theological view that sins of the fathers, and no doubt mothers, would be passed on.

Only in the last decade has evidence emerged that pregnant women should not drink, or drink much, for another reason: Heavy drinking may produce fetal abnormalities. The so-called "fetal alcohol syndrome" has been widely publicized and has absorbed most of the resources in time and money devoted to research about women and alcohol.

Exactly what is fetal alcohol syndrome?

It was first described with any precision in an obscure French gynecological journal in 1968 and again, with much more publicity, in 1973 by two Seattle pediatricians. These reports led to an almost evangelical search for the syndrome in children of drinking women. After many conferences, large amounts of money spent on research, and passionate disagreement, a rough consensus has emerged:

Fetal alcohol syndrome (FAS) has specific and nonspecific features. An FAS baby has a small head, short nose, thin upper lip, indistinct groove between upper lip and nose (called the philtrum), small eye openings, and flat cheeks. These facial features are also associated with maternal use of certain drugs, particularly those prescribed for epilepsy.

The nonspecific features of FAS are multitudinous. They include low birth weight, retarded growth as an infant and child, mental retardation, heart murmurs, birthmarks, hernias, and urinary tract abnormalities.

It now appears that between 30 and 50 percent of women who drink heavily during pregnancy have infants with one or more of these defects. The incidence of fetal abnormalities in children born of women who drink lightly or not at all during pregnancy appears to be about 5 to 10 percent. (Abnormality is used here in a broad sense, referring to anything from a birthmark to missing limbs.)

The term "possible fetal alcohol effects" has been suggested to describe these nonspecific abnormalities. It has become increasingly clear that the specific facial characteristics of FAS are fairly rare.

Some studies have failed to find FAS in any of the offspring of heavily drinking mothers. In one study of 12,000 deliveries it was reported that five of 204 children (2.5 percent) born of heavily drinking women had FAS. Even in these few cases, the examining doctors knew the mother had been a heavy drinker and may have been deliberately looking for FAS features.

One thing, however, seems clear. Women who have a history of heavy drinking during pregnancy have about a fifty-fifty chance of having a baby with some abnormality, and the abnormality most often reported is low birth weight.

There are two problems in interpreting this finding. One has to do with the definition of "heavy." The other refers to the fact that alcoholic women are often heavy smokers and often use drugs suspected of causing fetal abnormalities, tend to have poor nutrition, fall down a good deal, and generally have a lifestyle different from that of nonalcoholic women.

The definition of "heavy" remains elusive. It ranges from two or more drinks a day to eight or more, depending on the study. Even heavily drinking women drink erratically during pregnancy, ranging from binge drinking to abstinence and moderate drinking. Timing is crucial in fetal development. In a matter of seconds early in pregnancy a virus or drug can determine whether a limb is formed or not formed; birth size relates mainly to events occurring in the last two or three months of fetal development. It is not surprising that FAS has come to describe such a welter of observations.

How much a pregnant woman can drink without risk to the fetus is not known and perhaps never will be. Women are simply too diverse, and so are fetuses. To be on the safe side, pregnant women should not drink at all. Since for the first few weeks after conception many women do not know they are pregnant, this really means that women who are trying to conceive should abstain from alcohol altogether. In our society this is asking a lot.

There are some skeptics who question whether alcohol alone, in whatever amounts, produces specific fetal abnormalities in the same way as do German measles and thalidomide. It is difficult to separate the effects of alcohol from other factors, for example, smoking. At least 70 to 90 percent of alcoholic women smoke cigarettes. Women who smoke cigarettes tend to have small babies, whether they drink or not, so perhaps cigarette smoking and not alcohol is responsible for the low birth weight of babies born to alcoholic women.

Because nonsmoking alcoholic women are rare, this is a difficult hypothesis to test. However, at least three studies suggest that alcohol abuse and cigarette smoking contribute independently to the small size of the newborn. The best estimate at present is that alcohol abuse approximately doubles the risk of birth of a small infant, whether the mother smokes or not.

Heredity is another complicating factor when considering the effect of alcohol on pregnant women. Birth defects run in families. So does alcoholism. Is it possible that in some instances a common genetic predisposition explains both? There is little evidence for this one way or the other. However, some reports about twins suggest this explanation may not be so far-fetched. Sometimes one twin shows signs of FAS and the other does not; if alcohol is responsible, this should not happen.

There obviously is a great deal of uncertainty about FAS and how often it occurs or whether it occurs at all. To what extent should women modify their usual drinking patterns because of concern about FAS? Many women spontaneously reduce their alcohol intake during pregnancy because alcohol makes them ill. Should they reduce their intake to zero?

Investigators of FAS make the following recommendation:

> Observations of human babies and of experimentally treated animals have made it clear that a mother's heavy drinking can severely damage her unborn child. We do not know the exact amount or timing of drinking that causes these effects. We cannot say whether there is a safe amount of drinking or whether there is a safe time during pregnancy. We do know

that heavy drinking can be damaging. Women should therefore be especially cautious about drinking during pregnancy and when they are likely to be pregnant.

There are thousands of women in Alcoholics Anonymous and in alcoholism treatment facilities who have perfectly normal children, and this does not seem to get much attention. Many more children of problem-drinking mothers will have to be studied before fetal alcohol syndrome can be defined with certainty. Meanwhile, caution obviously is the wisest course, both for women who drink and for investigators who make premature and poorly substantiated claims.

Should women drink while breast-feeding? The alcohol drunk by the mother certainly appears in the milk. If the mother is intoxicated while breast feeding, the infant will also become intoxicated.

No one knows whether this will harm the infant. Newborns, like fetuses, are in a formative stage of development. One would expect them to be especially vulnerable to environmental insults like alcohol. Not drinking for a few hours preceding nursing, and during nursing, is undoubtedly the safest course.

THE ADDICTIVE CYCLE

The cause of alcoholism is unknown, but any ultimate explanation of alcoholism must account for two features of the illness: loss of control and relapse.

Loss of control refers to the alcoholic's inability to stop drinking once he starts. Relapse is the return to heavy drinking after a period of sobriety, and it is the great mystery of addictions. Why, after months or years of abstinence, does the smoker smoke again, the junkie shoot up again, the alcoholic fall off the wagon?

Here is an attempt to show how inherited and environmental factors may combine to produce loss of control and relapse.

Some people experience more pleasure, or glow, from alcohol than do others. How much pleasure a person gets from alcohol may be partly determined by heredity. The pleasure is short-lived and, in some people more than others, it is followed by a feeling of discomfort. The degree of discomfort may also be determined by heredity.

Alcoholics learn that the discomfort has a simple remedy: another drink. Thus the alcoholic drinks for two reasons: to achieve the pleasure and to relieve the discomfort. The same substance that produces the happy feeling also produces the unhappy feeling and is required both to restore the one and abolish the other. Nothing abolishes the unhappy feeling quite as effectively as the drug that produced it.

The unhappy feeling is called craving. To relieve craving the alcoholic will try anything—a chocolate bar, sex, tranquilizers, jogging, or prayer—but has learned that only alcohol gives complete and immediate relief. After a time, the alcoholic drinks more to overcome the unpleasant effects of alcohol than to attain the pleasant effects.

Some people are "born" to have higher highs and lower lows from alcohol than others. The highs and lows—the addictive cycle—may occur repeatedly in a single drinking session, and are followed by the monumental low known as hangover that comes the next day. The mind and body of the alcoholic "learns" that the lows can be banished by another drink. The learning is experienced physically as craving and psychologically as a preoccupation, literally an obsession, with having alcohol on hand at all times. Once the true binge drinker has started drinking, he often cannot stop until the high–low cycle has left him so exhausted that he must stop.

But why start in the first place? Why does the binge drinker, who actually has come to hate alcohol and hasn't drunk anything for months or years, one day start drinking again? It is not from ignorance; he knows what will happen. Why go through it *again*?

This brings up the problem of what is and isn't voluntary, which is a philosophical issue having to do with free will, and not a suitable topic for discussion here. Leaving aside free will, relapse can partly be explained by something called stimulus generalization.

The term refers to the fact that things remind people of other things. For the drinker, the hands of a watch pointing to 5 P.M. (the stimulus) may remind him that, for years, he always had a drink at 5 P.M. (the generalization), and so he drinks. Acts of drinking become embedded in a maze of reminders. Every drinker has his own reminders, but there are common themes. Food, sex, holidays, football games, fishing, travel—all have nothing intrinsically to do with drinking but all commonly become associated with drinking and are powerful reminders. Physical feelings become reminders: hunger, fatigue. Moods become reminders: nostalgia, sadness, elation. Anything, in short, can be a reminder and remain a reminder long after a person has stopped drinking.

Reminders may lead to relapse. One day, unexpectedly, the "recovered" alcoholic is flooded with reminders. It is 5 P.M. on Christmas Eve (which also is his birthday). The boss bawled him out and he missed lunch. His alimony check to his wife bounced, but he learns he just won the Irish Sweepstakes. Suddenly he has an incredible thirst. As he passes a pub, a strong wind blows him in the door—and a relapse occurs.

This is an extreme example. The relapse trigger may be subtle: "For any alcoholic," Mark Keller said,

> there may be several or a whole battery of critical cues or signals. By the rule of generalization, any critical cue can spread like the tentacles of a vine over a whole range of analogs, and this may account for the growing frequency of bouts, or for the development of a pattern of continuous inebriation. An exaggerated example is the man who goes out and gets drunk every time his mother-in-law gives him a certain walleyed look. After a while he has to get drunk whenever any woman gives him that look.

In either case, he probably will not be able to say later why he started drinking again. And maybe stimulus generalization is not the whole story. But it seems to explain a lot.

The idea of addictive cycles has been applied not only to alcoholism, but also to thrill-seeking, overeating, and love. The theory holds that every "addiction" eventually produces its opposite; pleasure turns to pain, and pain to pleasure.

Richard Solomon, a psychologist and leading proponent of this idea, believes that every event in life which has a strong effect also has an opposed process that fights it. At the start, he says, drugs are highly pleasant. You get a big "rush" and euphoria. But as tolerance builds up, the rush disappears and the threat and pain of withdrawal begin to take command. He compares the addictive cycle from drugs to a runner's high which, he says, is an example of pain giving way to pleasure. Parachute jumpers sometimes become extremely distressed when bad weather cancels their sport, reflecting, some believe, an "addiction" to jumping.

Some animal studies seem to support the idea of addictive cycles. In experiments measuring the distress calls of ducklings, for example, the offspring show far more distress when their mother is removed and returned at brief intervals than when she is removed for long periods of time. Frequent separations, in short, produce an addictive cycle in which the distress calls are the equivalent of withdrawal symptoms.

Such studies suggest that in its early stages any attachment is controlled mainly by pleasure, but late in the attachment the main control is the threat of separation and loneliness. Although the leap to human behavior is a long one, Solomon sees the same mechanisms at work. "The ecstasy and madness of the early love affair are going to disappear," he says, "and when they do, it means that a withdrawal symptom has to emerge if you are denied the presence of your partner."

According to the theory, the size of doses and the intervals between doses are crucial to addiction. The distress shrieks of ducklings are prolonged at one-minute intervals away from the mother, but not at two or five minutes. A rat fed a food pellet every sixty seconds shows withdrawal symptoms (agitated behavior, drinking too much water) after each morsel. But the symptoms disappear if the pellets are spaced several minutes apart. The implication: Proper timing of dosage prevents addiction.

Why do people keep eating when their stomachs are full? "Because we like to fight off withdrawal by redosing with a pleasurable taste," says Solomon. "The better the taste of the food, the harder the withdrawal." If you're watching your weight, it makes sense, he says, to eat tasty foods early in the meal and save bland ones for last, so the withdrawal will be easier. Better yet, eat only bland, uninteresting foods.

A weakness of the theory is that it doesn't adequately take into account the strength of addictive behavior. It's hard to believe that fat people would become thin merely by eating bland food at the end of meals. Still, who knows? Strategies for aborting the addictive cycle are being studied in various academic centers and may someday provide the basis for a rational therapy of the addictions, including alcoholism.

THE ALCOHOLIC PERSONALITY

There is none.

IS ALCOHOLISM A DISEASE?

My answer:

Diseases are something people see doctors for. Diseases are to doctors what groceries are to grocers: their specialty and their livelihood. What constitutes a disease has changed over historical time, but it usually indicates the presence of a known or suspected physical or psychological vulnerability that compromises the organism's ability to survive and function effectively in a particular environment.

Physicians are consulted about the problem of alcoholism and therefore alcoholism becomes, by this definition, a disease. It also fulfills a narrow definition of disease that requires the presence of a biological abnormality that leads to maladaptation. The evidence that alcoholism involves a biological vulnerability is just as strong, or perhaps stronger, than the evidence that hypertension or adult-onset diabetes involves a biological vulnerability. Hypertension and adult-onset diabetes both run in families. This is not conclusive evidence for "genetic" or "constitutional" factors, inasmuch as families often share the same dietary habits and lifestyles. However, it is widely assumed that both involve a biological susceptibility that often seems to require precipitating factors, such as overweight or excessive salt intake, to produce clinical disease.

The evidence that alcoholism is influenced by biological vulnerability comes mainly from adoption studies in which children of alcoholics, raised by nonalcoholic adoptive parents, still become alcoholic at a high rate: a rate three to five times higher than occurs in the families of nonalcoholics (see chapter 4). "Biological susceptibility" is not necessarily synonymous with genetic predisposition. It is conceivable that perhaps the mother was drinking during pregnancy and this in some fashion created the vulnerability. However, it seems more probable that genetic factors are involved. Also, like salt in hypertension and obesity in diabetes, habits, conditioning, learning, and availability of alcohol obviously are involved in the development of alcoholism in susceptible individuals.

To summarize: The evidence that alcoholism can properly be called a disease is just as strong for alcoholism as it is for many medical conditions universally regarded as diseases.

CHAPTER THREE

A Family Disease

"We are as days and have our parents for our yesterdays."
—Samuel Butler

"Drunkards beget drunkards."
—Plutarch

Alcoholism runs in families. This tendency was alluded to in the Bible. Aristotle and Plutarch remarked about it, and doctors and preachers of the nineteenth century were unanimous: Alcoholism ran in families *and* was inherited.

But this notion of inheritance was Lamarckian (referring to traits acquired through uterine exposure). If the mother took piano lessons, the child might have a musical talent. If the father drank, the sons might be drunkards.

By the first half of the twentieth century, this ancient piece of conventional wisdom had scientific underpinnings. Drunkard parents did indeed have drunkard children. They had them about four or five times more often than did parents who were not alcoholic. More than one hundred studies in the literature confirmed, beyond any doubt, that alcoholism was a family disease.

But in terms of scientific explanations, modern genetics had replaced Lamarck. For most psychologists and sociologists, "running in the family" no longer implied inheritance. It meant that children saw what their parents did and did the same, like learning French or voting Republican. Blue eyes were inherited; alcoholism was not.

By the 1970s, other questions were being raised. Did alcoholism *really* run in families like speaking French (environmental contribution) or did it run in families like blue eyes (genetic contribution)? How do you study such things? People like Donald Hebb said you could not. Writing about intelligence, Hebb warned against regarding intelligence as part heredity or part environment: "Each is *fully* necessary. . . . To ask how much heredity contributes to intelligence is like asking how much the width of a field contributes to its area."

Still, the familial nature of alcoholism was one of the few solid facts about alcoholism on which investigators could depend. Maybe separating nature from nurture was worth a try. Before discussing how to do this, we will first look at family studies and discuss their validity and meaning. Then we will focus on alcoholism that does not run in families—and illnesses associated with it.

THE FACTS ABOUT FAMILIES

To know whether a disorder occurs more often in some families than in others, it is necessary to know how often the disorder occurs in the general population. Estimates of alcoholism rates differ among countries and periods studied. The differences may reflect actual distinctions in prevalence, or they may arise from varying definitions of alcoholism and different approaches to obtaining information.

Even in small countries it is usually impossible to interview every member of the population. Therefore, a sample of the population must be identified, using scientific methods that are fairly recent and still being refined.

To be useful, the sample must be representative of the population. In other words, if you are studying left-handedness and 10 percent of your sample is left-handed, you must be able to assume that 10 percent of the general population also is left-handed. Sampling has been brought to its finest art in political polling. By sampling a tiny fraction of a population consisting of millions of voters, pollsters can predict the outcome of elections within several percentage points. It is harder to tell how valid or generalizable a sample is for other purposes. One has to assume that if methods valid in political forecasting are applied to comparable problems, they will remain valid, but unfortunately there is rarely anything comparable to an election to prove this.

There has been no nationwide attempt to determine the prevalence of alcoholism in the United States. Studies have been carried out in some areas, but none have been very satisfactory. One approach to ascertaining alcoholism rates is to base them on a formula that requires knowledge of cirrhosis rates. The premise is that if X percentage of alcoholics has cirrhosis, one can calculate the number of alcoholics from the number of people with cirrhosis. Many authorities question the usefulness of this formula. One reason is that cirrhosis rates are probably unreliable.

The point is this: The determination of base rates for a particular disorder is not easy, particularly when the disorder is as ill-defined as alcoholism. The best information about rates of alcoholism has come from other countries, notably Germany, Switzerland, Sweden, Denmark, and England. Studies in these countries have produced roughly equivalent findings. The expectancy rate for alcoholism among men appears to be about 3 to 5 percent. The rate for women ranges from 0.1 to 1 percent. The first study was conducted in 1928 and the most recent in 1948. Whether the rates would be higher or lower now is not known. It may be, for example, that the rate for women has risen with changing social mores. Nevertheless, together, these studies provide the best basis for estimating the prevalence of alcoholism in Western countries. For alcoholism to "run in families," susceptible families must have rates of alcoholism considerably higher than 5 percent in men and 1 percent in women. And that is what has been found. Without exception, every family study of alcoholism has shown much higher rates of alcoholism among the relatives of alcoholics than occur in the general population.

Family studies

One of the first and largest family studies was carried out in Germany in 1929. Nearly a thousand male alcoholics and 166 female alcoholics were examined. Alcoholism occurred in half of the fathers, 6 percent of the mothers, 30 percent of the brothers, and 3 percent of the sisters. Later studies have shown somewhat lower rates of alcoholism in the fathers and brothers of alcoholics, but in none are less than 25 percent of the fathers and brothers alcoholic, a rate at least five times greater than the maximum estimate for the male population.

Apparently it does not matter whether male or female alcoholics are studied; in either case the rates of alcoholism among their male relatives range from 25 to 50 percent and among their female relatives from 5 to 8 percent. In short, both male and female relatives have rates of alcoholism at least fivefold what would be expected in the general population. The rate appears to be highest with hospitalized alcoholics, roughly half of whom come from families where one or both of the parents abused alcohol.

One interesting question regards the specificity of the susceptibility to addiction. In other words, are relatives of alcoholics prone only to alcoholism or are they prone to drug abuse of all kinds, with the choice depending on what is available?

A 1933 study in Germany addressed itself to this question. The investigator compared the relatives of alcoholics with the relatives of morphine addicts. Alcoholism was found in one quarter of the brothers of alcoholics and in only 6 percent of the brothers of morphine addicts. Alcoholism occurred in half of the fathers of alcoholics and, again, in only 6 percent of the fathers of morphine addicts. Conversely, among relatives of morphine addicts, morphine addiction occurred more frequently than did alcoholism. Based on this one study, it appears that both alcoholism and morphine addiction run pure to type in families, with little overlap. Chapter 5 presents further evidence supporting this view.

However, there is evidence contradicting the notion of drug-specific proneness to addiction. For example, many heroin addicts drink to excess, particularly when heroin is not available to them. Furthermore, alcoholics are notoriously "addicted" to caffeine and nicotine, and a substantial proportion of alcoholics use other drugs—particularly tranquilizers—although it is not clear whether they abuse these drugs.

Furthermore, cultural factors and availability are important in determining drug choice. In a country where alcohol was the only available drug, it would not be surprising if only alcoholism was transmitted from generation to generation. LSD was first synthesized in 1950; before then there was a zero rate of LSD abuse in the families of alcoholics. On the other hand, barbiturates, chloral hydrate, paraldehyde, and the bromides have been available for many years, as well as morphine and its derivatives, and the question of whether "polydrug" abuse, to use a currently fashionable term, also runs in families remains an open one. One of the obvious difficulties in answering the question is continued disagreement about the definition of "abuse."

Investigator bias

"Seek and ye shall find," says the Bible, and it is possible that if you expect to find alcoholism in the families of alcoholics, you will indeed find it. In the family studies summarized above, there is no way of knowing what the investigators' expectations were, but it would be reasonable to assume that they would not be conducting the study unless they expected "positive" results, namely, a finding of an increased prevalence of alcoholism in the families of alcoholics.

In other words, the investigators may have had a bias. They may have seen alcoholism where it didn't exist. If an alcoholic told them that "Dad drank," they may have categorized Dad as an alcoholic more readily than if a nonalcoholic told them that his father drank. Where only secondhand information about families is available, biased interpretations would appear particularly likely. Rarely are the family members personally interviewed, and even when they are, investigators might be biased, knowing they are interviewing families of alcoholics.

The ideal study would be "blind," one in which family members of alcoholics are interviewed by interviewers who do not know whether they are related to alcoholics. Few studies take this precaution.

What the studies mean

Meanwhile, most students of alcoholism would view the above as quibbling. The fact is that for seventy-five years or more, studies have consistently reported much higher rates of alcoholism in the families of alcoholics than in the general population. This finding has been so consistent that hardly anyone at present doubts its validity, yet we are still searching for an answer to the question: What does the family do to increase the risk? Here are some possibilities.

1. The family provides genes that make a person vulnerable to alcohol, so that use leads to abuse.
2. The parents are "bad" parents. They frustrate the child, making him feel anxious in later life. He drinks to feel less anxious.
3. The family "teaches" the child to drink. The boy sees his father drink and follows his example. He learns from observation that certain problems can be solved by drinking: fatigue, depression, shyness.
4. The values of the family are transmitted to the child, and these values promote drinking. For example, young men who cannot cry because of the machismo they learned from their fathers may find that alcohol provides consolation for occasions when crying would be consoling.

These are some of the many theories explaining why one in twelve or one in fifteen drinkers is alcoholic. They will be discussed in detail in chapters 5 and 6. Meanwhile, alcoholism research has taken a new turn. Researchers are now studying not only alcoholics with alcoholism in the family but also alcoholics without alcoholism in the family.

ARE THERE TWO TYPES OF ALCOHOLISM?

Several years ago, investigators made what they thought was a new observation: Some alcoholism ran in families and some did not. Were there differences between alcoholics with alcoholism in the family and alcoholics without? It turned out there were.

First, however, it should be noted that the observation that alcoholism comes in two types—familial and nonfamilial—is *not* new. It simply has been long ignored.

The turn-of-the century novelist Jack London wrote a book about his own drinking called *John Barleycorn.* It is a classic, must reading for anyone interested in alcoholism. There, on page 2, he introduces the concept of "hereditary" alcoholism versus "acquired" alcoholism.

The insight had come to him on election day. He had been drinking and, to the amazement of his wife, voted for the suffrage amendment. But let London tell it.

> I was lighted up. In my brain every thought was at home. Every thought, in its little cell, crouched ready-dressed at the door, like prisoners at midnight waiting a jail-break. And every thought was a vision, bright-imaged, sharp-cut, unmistakable. My brain was illuminated by the clear, white light of alcohol. John Barleycorn was on a truth-telling rampage, giving away the choicest secrets on himself. And I was his spokesman. . . .
>
> I outlined my life to Charmian [his wife], and expounded the makeup of my constitution. *I was no hereditary alcoholic. I had been born with no organic, chemical predisposition toward alcohol* [italics added]. Alcohol was an acquired taste. It had

been painfully acquired. Alcohol had been a dreadfully repugnant thing—more nauseous than any physic. Even now I did not like the taste of it. I drank it only for its "kick." And from the age of five to that of twenty-five, I had not learned to care for its kick. Twenty years of unwilling apprenticeship had been required to make my system rebelliously tolerant of alcohol, to make me, in the heart and the deeps of me, desirous of alcohol.

I sketched my first contacts with alcohol, told of my first intoxications and revulsions, and pointed out that always one thing in the end had won me over—namely, the accessibility of alcohol. Not only had it always been accessible, but every interest of my developing life had drawn me to it. A newsboy on the streets, a sailor, a miner, a wanderer in far lands, always where men came together to exchange ideas, to laugh and boast and dare, to relax, to forget the dull toil of tiresome nights and days, always they came together over alcohol. The saloon was the place of congregation. Men gathered to it as primitive men gathered about the fire of the squatting-place or the fire at the mouth of the cave.

. . . As a youth, by way of the saloon I had escaped from the narrowness of women's influence into the wide free world of men. All ways led to the saloon. The thousand roads of romance and adventure drew together in the saloon, and thence led out and on over the world.

The point is that it is the accessibility of alcohol that has given me my taste for alcohol. I did not care for it. I used to laugh at it. Yet here I am, at the last, possessed with the drinker's desire. It took twenty years to implant that desire; and for ten years more that desire has grown. And the effect of satisfying that desire is anything but good. Temperamentally I am wholesome-hearted and merry. Yet when I walk with John Barleycorn I suffer all the damnation of intellectual pessimism.

. . . That is why I voted for the amendment today. I read back in my life and saw how the accessibility of alcohol had given me the taste for it. You see, comparatively few alcoholics are born in a generation. *And by alcoholic I mean a man whose chemistry craves alcohol and drives him resistlessly to it* [italics added]. The great majority of habitual drinkers are born not only without desire for alcohol but with actual repugnance toward it. Not the first, nor the twentieth, nor the hundredth drink, succeeded in giving them the liking. But they learned, just as men learn to smoke; though it is far easier to learn to smoke than to learn to drink. They learned because alcohol was so accessible. The women know the game. They pay for it—the wives and sisters and mothers. And when they come to vote they will vote for prohibition. And the best of it is that there will be no hardship worked on the coming generation. Not having access to alcohol, not being predisposed toward alcohol, it will never miss alcohol. It will mean life more abundant for the manhood of the young boys born and growing up—ay, and life more abundant for the young girls born and growing up to share the lives of the young men.

"Why not write all this up for the sake of the young men and women coming?" Charmian asked. "Why not write it so as to help the wives and sisters and mothers to the way they should vote?"

London did write it up. *John Barleycorn* was a 1912 best-seller and was serialized in the *Saturday Evening Post,* the most widely read periodical of the day. Many gave London, the country's most popular writer, much credit for the passage of prohibition in 1919. His message was simple: Let the women vote and they will vote out liquor. And so they did.

But his other message—that there are two types of alcoholics, hereditary and nonhereditary—was overlooked throughout the twentieth century until the 1980s. Now it has been rediscovered. A growing number of studies have consistently found differences between familial (hereditary) and nonfamilial alcoholics, among them the following:

1. Hereditary (familial) alcoholics are younger when they become addicted.
2. They become addicted in a shorter time—sometimes, literally, with the first drink.
3. The addiction is particularly severe.

The differences are particularly impressive when there is a strong family history of alcoholism. Indeed, most alcoholics who have one alcoholic family member almost always have two or more.

Other differences are reported. Concerning one difference there is controversy. Some scientists find that familial alcoholics are more often antisocial than are nonfamilial alcoholics. Other writers fail to find this. The latter view familial alcoholism as synonymous with "primary" alcoholism. According to this view, the familial alcoholic has one problem: He or she drinks too much. If the person stops drinking, he or she is no more or no less likely to have another psychiatric condition than people in the general population.

Nonfamilial alcoholics, by contrast, can be called "secondary" alcoholics. They often have psychiatric illnesses other than alcoholism. Perhaps, indeed, the alcoholism is caused by—or is "secondary" to—the other illness.

The illnesses that nonfamilial or secondary alcoholics are most likely to have will be discussed a little later. First, I'll describe some of the other differences found between familial and nonfamilial alcoholics.

Familial alcoholics more often have a history of hyperactivity and behavior problems as schoolchildren. This may be one reason why some investigators find that familial alcoholics also show a high rate of antisocial behavior. So-called antisocial personalities are formed early in life, the person committing antisocial acts as far back as childhood. There is a distinct overlap between the early years of a future criminal and the hyperactive child syndrome with conduct disorders. The controversy revolves around whether the antisocial behavior seen in adulthood in alcoholics is a consequence of the drinking, independent of the drinking, or merely a late manifestation of the hyperactive child syndrome.

Familial alcoholics do less well on psychological tests for brain damage than do nonfamilial alcoholics, even when the alcoholism is equally severe in both groups.

Familial alcoholics respond less well to treatment. "Cure" consists of total sobriety. There is some evidence that nonfamilial alcoholics may be "trained" to control their drinking to some extent, but therapists who work in this field are not optimistic that this can be accomplished with familial alcoholics.

Familial alcoholics are more often left-handed. This has been found in several studies, most from the same institution.

Alcoholism runs in families, and there is evidence (reported on in chapter 5) that it runs in families even when the children are separated from the biological alcoholic parents and raised by nonalcoholic adoptive parents. The latter finding is widely interpreted as the strongest evidence available that alcoholism is influenced by genetic factors.

However, some of the differences found between familial and nonfamilial alcoholics suggest a different explanation. Perhaps the mothers drank during pregnancy, and this explains why some of the children have hyperactive syndromes and antisocial behavior. All of these differences can be lumped under the category of minimal brain damage. Perhaps the tendency to become alcoholic reflects this minimal brain damage resulting from the mother's drinking during pregnancy. In other words, perhaps alcoholism itself is a form of fetal alcohol syndrome!

In the adoption studies described later, little is known about the mother's drinking habits during pregnancy. It is known that alcoholics tend to marry alcoholics, and possibly the mothers were drinking during pregnancy, but there is no direct evidence of this. Also, most children of alcoholics do not have any of the signs listed above—hyperactive syndrome, conduct disorders, antisocial behavior, or abnormal brain findings.

In any case, the effects of maternal drinking during pregnancy offer an alternative explanation to the genetic hypothesis for the development of alcoholism. Most behavioral geneticists involved in the study of alcoholism reject the possibility as remote, but it exists.

To summarize: A current theory holds that there are two types of alcoholism; one runs in families and the other does not. The first could be considered primary in that it is not associated with other psychiatric conditions (except, perhaps, antisocial personality). The nonfamilial type is often associated with other psychiatric illnesses and indeed may be caused by other illnesses. The practical implication of this for therapists who work with alcoholics is that, once having determined whether the patient has a family history of alcoholism, the therapist can then predict with some accuracy whether or not other psychiatric conditions exist. Presumably, the stronger the family history of alcoholism, the more likely this prediction will turn out to be true. In any case, it is a prediction that can be tested in the office practice of anyone working with people with alcohol problems.

Here are some of the conditions one might likely encounter in nonfamilial alcoholics:

1. Depression and/or mania
2. Anxiety disorders
3. Antisocial behavior
4. Essential tremor
5. Early dementia
6. Polydrug abuse
7. Situational stress

When other conditions are diagnosed, they may or may not have a causal relationship to the drinking. When the depressed patient drinks, should the depression be viewed as cause or consequence of the drinking? When the phobic drinks in order to give a public speech or to go to a party, should it be viewed as self-treatment or self-indulgence? Antisocial personalities are notoriously heavy drinkers, but how much of the antisocial behavior is a consequence of the drinking?

I'll discuss each condition briefly, beginning with depression. A connection between depression and alcoholism has long been suspected, and it is important to review the evidence for and against the association.

Alcoholism and depression

> Melancholy is at the bottom of everything, just as at the end of all rivers is the sea. . . . Can it be otherwise in a world where nothing lasts, where all we have loved or shall love must die? Is death, then, the secret of life? The gloom of an eternal mourning enwraps every serious and thoughtful soul, as night enwraps the universe.
>
> —Henri-Frederick Amiel

> And malt does more than Milton can
> To justify God's ways to man.
>
> —A. E. Housman

God's ways have struck many people, including Amiel, as profoundly depressing. What is a good cure for depression? Alcohol, Housman says, and through the ages people have agreed with him.

But is it so? Does alcohol relieve depression? Will it relieve serious depression, the kind psychiatrists see?

More important, does depression cause alcoholism? Are alcoholics really just depressed people who drink to feel better (and better, and better) until drinking becomes uncontrolled, a habit with its own propulsion, progressing independently of the depressed feelings that caused it?

Some believe it. What is the evidence? Here are the questions that need to be asked:

How many alcoholics have manic-depressive disease?

Manic-depressive disease is a condition manifested by episodes of depression that are prolonged, disabling, and that often require treatment. These may be interspersed with episodes of mania.

Many alcoholics become depressed. Their depressions resemble the depressions seen in manic-depressive disease. The alcoholics become irritable and can't sleep. They feel melancholy and sad. They experience feelings of guilt and remorse. They lose interest in life and contemplate suicide.

Indeed, suicide is a common outcome of alcoholism. Except for manic-depressives, alcoholics commit suicide more than any other group. One reason to believe alcoholism and manic-depressive disease are related is that, in Western countries, most people who commit suicide have one or the other illness. Many psychoanalysts believe that alcoholism and manic-depressive disease have the same origin; victims of both illnesses are orally fixated and, instead of feeling angry toward others, feel angry toward themselves. Aggression directed toward one's self is experienced as a feeling of depression. The ultimate act of self-aggression is suicide. Alcoholism has been called "slow suicide."

This is an interesting theory, but difficult to prove scientifically. Although the depression experienced by alcoholics resembles manic-depressive disease, there is a difference: When the alcoholic stops drinking, the depression often goes away. Alcohol is a toxin. In large amounts it produces depression, anxiety, irritability. The alcoholic feels guilty and for good reason: He has botched up his life and he knows it.

Based on most studies it appears that alcoholism causes depression more often than depression causes alcoholism. As noted, the cure for alcohol-induced depression is not antidepressant pills or electroconvulsive therapy but abstinence. (The toxic effects, incidentally, may last for weeks or months after a person stops drinking. Often several months pass before sleep becomes normal again.)

Chapter 2 discussed the difference between women alcoholics and men alcoholics. One difference is that women alcoholics more often have depressions preceding the onset of heavy drinking. In women more than men, a case can be made that alcoholism is sometimes a manifestation of depression. As a group, women start drinking heavily at an older age than men. In people who are drinking heavily it is hard to tell whether the depression or alcoholism came first. Since women become alcoholic at an older age than men, there is more opportunity for depressions to occur before heavy drinking occurs. Depression may promote heavy drinking in men as often as it appears to in women, but because men start drinking heavily at an earlier age, it is impossible to determine whether the depressions are a cause or consequence of the drinking.

How many manic-depressives are alcoholic?

This sounds like the same question as the preceding one, but isn't. If you study a group of manic-depressive patients, how many are alcoholic? Such studies have been conducted, with conflicting results. In one study, one third of manic-depressives were alcoholic. In another study, 8 percent were alcoholic. Eight percent is not much higher than the prevalence of manic-depressive disease (as broadly defined here) in the general population. Which study is correct? Nobody knows.

Do manic-depressives drink more when they are depressed?

Nonalcoholic manic-depressives have no established drinking pattern. Probably more manic-depressives cut down on their drinking when depressed than increase their drinking.

One investigator gave alcohol to hospitalized manic-depressives. A small amount of alcohol improved their mood. A large amount made them more depressed. The same is true of alcoholics. Bring alcoholics into an experimental ward, give them alcohol for a long period, and instead of feeling happy they feel miserable. It turns out alcohol is a rather weak euphoriant compared, for example, to cocaine and amphetamines. The hedonistic explanation for alcoholism, contrary to popular opinion, has little support in science.

After giving alcohol to many psychiatric patients, including manic-depressives and alcoholics, Dr. Demmie Mayfield at Kansas University, who conducted the above study, reached the following conclusions:

- If you feel bad, drinking will make you feel a lot better.
- If you drink a lot, it will make you feel bad.
- Feeling bad from drinking a lot does not seem to make people choose to stop. Feeling a lot better from drinking does not seem to encourage people to continue drinking.

Do manic-depressives drink more when they are manic?

Definitely. Many manics who ordinarily drink little or nothing start drinking heavily when manic. The explanation is not clear, but it seems that alcohol has almost a specific ameliorative effect on the manic mood. Manics often feel *too* high, uncomfortably high, and alcohol seems to reduce the unpleasant effects of the mania while amplifying the pleasant ones. Lithium, an effective drug for mania, has been given to alcoholics and seems to reduce the frequency of relapse, particularly if the alcoholics are depressed. More studies are required before it can be known definitely whether lithium is helpful for alcoholism.

Do alcoholism and manic-depressive disease run in the same families?

If they did run in the same families, this would suggest a common genetic predisposition. There is evidence that both alcoholism and manic-depressive disease are influenced by heredity, but is it the *same* heredity?

The evidence is mixed. Some studies show an increase of depression in the families of alcoholics and of alcoholism in the families of manic-depressives. Other studies fail to show this. Those that show an overlap have resulted in a concept called depressive spectrum disorder. In this still hypothetical condition, alcoholism occurs on the male side of families and depression on the female side. Based on adoption studies, there is some evidence that daughters of alcoholics suffer depressions when they are raised by their alcoholic parents but not when they are separated from their alcoholic parents and raised by nonalcoholic adoptive parents. This suggests that the depression seen in female relatives of alcoholics may be less influenced by heredity than by the environmental circumstance of being raised by an alcoholic parent.

Does antidepressant medication relieve alcoholism?
If it did, of course, this would suggest that alcoholism was caused by depression. The best studies, however, do not indicate that antidepressant medication is useful for alcoholism.

To summarize: Alcohol may make depressed people feel a little better, but not for long. It may, indeed, make them more depressed. The evidence that depression causes alcoholism is weak. Women, more than men, become alcoholic following or during a depression, but the causal connection is still not established. Even if depression doesn't cause alcoholism, alcoholism certainly is depressing and most drinking alcoholics are depressed a good deal of the time.

Anxiety disorders

Alcoholics have always been considered anxious people, but not until recently were they known to be particularly subject to anxiety disorders. As Vaillant found in his follow-up study (see chapter 2), people who later became alcoholic apparently were not particularly anxious in their childhood and teens. Vaillant did not address whether they had other psychiatric illnesses between the initial study and the forty-year follow-up.

According to several recent studies, about one third of alcoholics give a history of having anxiety disorders in their late teens or early twenties that might have been a reason for drinking. In men, the most common problem consisted of social phobias: fear of being in social circumstances where they might be judged and criticized. Women more often gave a history of agoraphobia: refusal to leave the house because of fear of having a spontaneous panic attack. Presumably, some alcoholics began drinking in order to overcome their phobias. Over a period of time the drinking increased until they acquired a chemical dependence on alcohol. With a few drinks they could overcome their phobias but, sober, they were just as phobic as ever.

Antisocial personality (sociopathy)

Sociopaths—people who from childhood flout society's laws and mores—constantly get in trouble and some of the trouble is connected with drinking. This group also uses street drugs and engages in a variety of risky behavior, such as robbing banks. It is usually impossible to tell whether crimes committed by sociopaths would have happened if they were not drinking. It is known, however, that when sociopaths stop drinking (if they do) they seem to commit fewer crimes. However, the age when they stop drinking is also the age when criminals stop committing crimes. Therefore, the role of alcohol is often uncertain.

Most people in penitentiaries are sociopaths and about half are very heavy drinkers. As discussed earlier, there is controversy about whether familial alcoholics are more likely to be sociopathic than nonfamilial alcoholics, a controversy still not resolved.

Essential tremor

Essential tremor is a familial condition in which a person has a fine tremor of the hands, particularly under stress. It can be embarrassing in situations where the tremor can be observed by others. Just lifting a coffee cup can be an occasion for embarrassment.

There is a marvelous cure for essential tremor: a drink or two. Alcohol makes the tremor disappear entirely. Victims often learn this and drink in situations where the tremor would be an embarrassment. Just as drinking to relieve phobic anxiety may intensify to the point of being a problem in itself, drinking to relieve a tremor may lead to another type of tremor—hangover shakiness. How many people become alcoholic through attempts to treat a tremor with alcohol is not known, but some do.

Dementia

People who can tolerate alcohol without problems for most of their lives may find, on reaching old age, that even a modest amount of alcohol affects their behavior. Alcoholism very rarely begins after mid-life. When an elderly person seems intoxicated and does not have a prior history of heavy drinking, the explanation may be early dementia. People who have experienced any form of brain damage seem sensitive to the effects of alcohol. In this category are former Golden Glove boxers and people with a recent history of encephalitis. When an elderly person becomes intoxicated while drinking it should not be assumed automatically that he or she is consuming large amounts. A drink or two may suffice. Also, a medical evaluation for dementia may be indicated.

Polydrug abuse

The 1960s introduced a new species of drug abuse into the United States called "polydrug abuse." The polydrug abuser is almost always a young person, often sociopathic and apparently willing to take any drug available in all combinations. At the same time the polydrug abuser may be intoxicated on hallucinogens, cocaine, alcohol, marijuana, plus other substances not easily identified. The person may be particularly dependent upon alcohol because of its availability, but the multiple use of drugs, not just alcohol, dominates the picture.

Many older alcoholics use other drugs and even use them over long periods, but their substance of choice is alcohol, and the use of tranquilizers or other drugs usually reflects a desire to stop drinking, get some sleep, clean up one's act. These alcoholics are very different from the polydrug user, who is a relatively new phenomenon in America and perhaps the toughest of all cases to treat.

Situational stress

Some moderate drinkers when stressed heavily increase their drinking, even to the point of having withdrawal symptoms.

Divorce in modern-day America is a stress particularly identified with heavy drinking. Singles bars are not only places to meet members of the other sex (or same sex) but also *bars*, places where alcohol is served. "Candy is dandy but liquor is quicker," Ogden Nash said, and the singles bar serves several functions, among them seduction, and reduction of the stresses associated with divorce.

To summarize again:

Alcoholism runs in families. It runs in families even when the children are separated from the alcoholic parents and raised by adoptive parents. But there are also alcoholics with no alcoholism in the family.

This chapter concludes with a hypothesis: Alcoholics with a strong family history of alcoholism are primary alcoholics. Their alcoholism appears unrelated to other psychiatric conditions. Cure consists of total and permanent abstinence. The nonfamilial alcoholic more often has secondary alcoholism, meaning that another psychiatric illness is likely to exist.

The hypothesis advanced in this chapter is that therapists, by learning whether their patient does or does not have alcoholism in the family, can predict with fair assurance whether another psychiatric condition exists, possibly one that is treatable. The hypothesis has some data to support it but not enough to make the hypothesis a fact. It can easily be tested by any clinician who treats alcoholism.

CHAPTER FOUR

Nature and Nurture

> The fish is in the water and the water is in the fish.
> —Arthur Miller

Many things run in families: diabetes, speaking French, musical talent, voting Republican, money, longevity, and cake recipes. But this does not mean that these things are genetic. Speaking French certainly is not, although fluency in French may be. Musical talent may be genetic, but reading music is not. Voting Republican may be a case of genes plus environment. Some Republicans may be born Republican, that is, born with personality traits that predispose to Republicanism.

Even diseases like diabetes, in which a genetic factor is assumed, may be influenced by environment. Perhaps children of diabetics, by family custom, are fed large amounts of sugar, which increase their chance of becoming diabetic. It is unlikely, of course.

As Brendan Gill pointed out, money grows on trees—family trees. The fact that longevity runs in families has led to a medical aphorism—"If you want a long life, have long-lived ancestors"—without answering the question whether genes or environment are mainly responsible.

Separating heredity and environment is not easy. The providers of our genes usually provide an important part of our environment. A few years ago the Danish psychiatrist Erik Strömgren asked, "Is it at all possible to distinguish between genetic and environmental factors in the causation of alcoholism?" The problem, he noted, had been the object of endless discussions. He answered by giving a case history.

> During a psychiatric examination the patient states that he is drinking because he is so nervous and needs help against the anxiety and feeling of inferiority. This might indicate that his alcoholism is caused by inborn components of his personality.
>
> But then the patient adds that he was not like this before, it started after he had been drinking for some years. Maybe, then, his nervousness has arisen on a toxic basis. . . . But now the patient informs us that alcohol has always had an abnormal action on him, his behavior is definitely abnormal when he is drinking. Maybe there is a constitutional abnormal reaction to alcohol.
>
> Further on the patient states that difficulties in his marriage made him start drinking; again an environmental factor. His wife, however, is of the opinion that the problems in the marriage were caused by the fact that the husband has always been a peculiar person, even before he started drinking, extremely sensitive and jealous, with difficulties of contact except when he was a little intoxicated. Constitution again? But then we find out that his childhood environment was really very bad and might be responsible for his disharmonious character.
>
> Finally, it turns out that the personality of his father was of exactly the same type as his own personality and that a number of similar types are found in the family. Perhaps genetics take the lead again.
>
> We have all met cases like this. They are no exception and such experiences make you quite humble with regard to making positive statements concerning the causation of alcoholism in the individual patient.

The humility is of fairly recent origin. In the late nineteenth and the early twentieth century, much attention was given to the problem of determining how much was inherited in human behavior, as opposed to how much was learned, with strong positions taken on both sides. Concern with the "nature–nurture problem" declined in the 1930s and had all but disappeared by World War II. Until recent years there was little research interest in heredity as a determinant of alcohol use and alcoholism. In part this came from an increasing awareness of the complex interrelationships of genetic and environmental factors, convincing many scientists that nature and nurture could never be unraveled.

This is a popular view today. Peter Medawar, the Nobel laureate biologist, says there is no *general* solution of the problem of estimating the size of the relative contributions of nature and nurture. "The reason is that the size of the contribution made by nature is itself a function of nurture." (He uses the word *function* in its mathematical sense, i.e., x is a function of y when the value of x varies depending on the value of y.)

> If someone constitutionally lacks the ability to synthesize an essential dietary substance, say X, then the contribution made by heredity to the difference between himself and his fellow men will depend on the environment in which they live. If X is abundant in the food he normally has access to, his inborn disability will put him at no disadvantage and may not be recognized at all; but if X is in short supply or lacking, then he will become ill or die. The same reasoning applies to other, much more complicated examples. If people live a simple pastoral life that makes little demand on their resourcefulness and ingenuity, inherited differences of intellectual capability may not make much difference to their behavior; but it is far otherwise if they live a difficult and intellectually demanding life. How often has it not been said that the stress of modern living raises the threshold of competence below which people can no longer keep up or make the grade? This is not to deny that some differences between us are for all practical purposes wholly genetic, wholly inborn. A person's blood group is described as "inborn" not just because it is specified by his genetic makeup, but because (with certain rare and known exceptions) there is no environment capable of supporting life

> in which that specification will not be carried out. Most differences between us are determined both by nature and by nurture, and their contributions are not fixed, but vary in dependence on each other.

The fact remains that alcoholism runs in families. Any attempt to understand the development of alcoholism must account for this, by now, unchallengeable fact. *Why* does alcoholism run in families? *How* does it run in families? It is almost impossible to approach these questions without asking: What is inborn? What comes later? Nature has provided only a few ways even to begin to distinguish between the two.

TERMS

Terms like *genetic* and *heredity* are ambiguous, and definitions are in order. A gene is a unit of heredity. It is a specific set of chemical instructions stored in every cell which call for specific chemical events, resulting in specific inherited traits. Genetic, in its narrow sense, means dictated by genes. Unfortunately, few inherited traits are produced by single genes (they include blood types and a few illnesses), but much of our modern knowledge of genetics is based on the study of these traits.

Colorblindness is an example of a specific inherited trait. Women with the gene for colorblindness are not colorblind themselves but, on the average, half of their sons are. When the statement is made that colorblindness is inherited, no one questions it: In families with the colorblindness gene the number of males afflicted can be predicted with mathematical precision. Environmental factors presumably have no influence.

Most inherited traits and illnesses are caused by more than one gene, and this presents difficulties. Brown-haired men often have brown-haired sons, and diabetic parents often have diabetic children, but you cannot predict with mathematical certainty how often. "Polygenic" traits and illnesses encourage controversy about the relative importance of genes and environment.

For example, from the moment the male sperm cell fertilizes the female egg, chemical processes dictated by genes are influenced by environment: nutrients, toxins, hormones, living organisms such as bacteria and viruses, all as unrelated genetically to a person as his next-door neighbor. Their combined influence on genetically dictated processes is so complicated as to be incalculable.

Whenever a woman drinks alcohol while pregnant, alcohol becomes part of the chemical environment of the embryo. Is it harmful to the embryo? Apparently so, in sufficient amounts. Various drugs, not to mention the German-measles virus, have devastating effects on the embryo if they enter its environment at critical periods of development.

Thus genetic and heredity can be used in a narrow sense, referring to traits transmitted solely by genes, or in a broad sense, referring to traits possibly transmitted by genes but also possibly caused or modified by prenatal environmental influences. When people say "genetic" or "inherited," often they are really referring to everything that happens prenatally; when they say "environment," they are usually referring to things that happen postnatally. In truth, both growth and regression in all stages from the embryonic to senile decay are to some degree under genetic control. Many genetically determined conditions—baldness in men, for example—do not usually become manifest until well on in adult life. And embryonic development is influenced by the intrauterine environment from the moment of conception.

Finally, there are a number of well-understood genetic disorders, including certain blood diseases, which do not manifest themselves unless triggered by environmental agents, such as drugs. The person with the disease never knows he has it until he takes a certain drug; in a practical sense, he does not have the disease until its symptoms appear. In alcoholism the situation is similar. Even if it could be proved that alcoholism is inherited, an environmental trigger—namely, alcohol—has to be introduced in order for the disorder to develop.

Basically, there are three ways to separate heredity from environment.

TWINS

A century ago, Sir Francis Galton proposed that twins are an "experiment in nature" ideally suited for separating "learned" behavior from inherited or "unlearned" behavior.

It is science's good fortune that twins come in two forms—identical and fraternal. Identical twins have the same genes; fraternal twins are siblings sharing some but not all genes who happen to have been born at the same time. If one *identical* twin has a simple genetic trait, such as colorblindness, the chance of the other twin being colorblind is 100 percent. If one *fraternal* twin is colorblind, the chance that the other twin will be colorblind is 50 percent if a boy and 0 percent if a girl; in other words, the probabilities are the same as they would be if they were brothers and sisters, which is exactly what they are. (The sex differences occur in sex-linked traits, such as colorblindness, not in so-called autosomal traits.)

Twin studies also are valuable in studying polygenic traits and illnesses. One compares identical and fraternal twins with respect to concordance for the condition being studied (concordant means that both twins have the condition). If identical twins are concordant for alcoholism more often than fraternal twins, heredity may be a factor.

Schizophrenia, depression, criminality, and alcoholism are conditions that run in families and have been studied by the twin method. Most studies show that identical twins are more concordant for these conditions than are fraternal twins, suggesting a genetic factor. But there are problems.

For one thing, twin studies vary markedly. Of the dozen twin studies of schizophrenia, concordance rates for identical twins range from 85 percent to 6 percent, and for fraternal twins from 19 percent to 5 percent. True, with one exception, all of the studies show higher concordance rates in identical twins than in fraternal twins, but why the discrepancies? There are several possibilities:

1. *Definitional.* There is even more disagreement about diagnosing schizophrenia than about diagnosing alcoholism.
2. *Bias.* Believers in a genetic basis for schizophrenia may unknowingly overdiagnose schizophrenia in identical twin brothers of schizophrenics and underdiagnose schizophrenia in fraternal brothers.
3. *False assumptions about twins.* It is known that identical twins share the same genes and fraternal twins genetically are no more or less similar than are brothers and sisters who are not twins. It is *assumed* that identical and fraternal twins are treated in the same way by important persons in their environment—parents, siblings, teachers, and so forth.

   This may not be true. A person's appearance, for example, influences people's behavior toward him; individuals who look alike may be treated alike, and identical twins look more alike than fraternal twins. Also, identical twins tend to be more comradely than fraternal twins and are more likely to have similar life experiences. They live together longer and more often have a comparable educational, vocational, and marital status. These similar life experiences may reflect common genes, but it is hard to prove. Family and peers may respond differently to identical twins than to fraternal twins, and this may partly explain why identical twins are more alike.

ADOPTEES

Not all children are raised by the people who provide their genes, their "real" parents. Some are adopted by foster parents. When the adoption occurs in early infancy; when the foster parents are unrelated to the genetic parents; when the adopted child has no contact with the genetic parents or other blood relatives—then the conditions are suitable for a nature–nurture study.

Assume that one of the genetic parents has disease X. The

foster parents do not have disease X. Will the children of parents with disease X also develop disease X even though separated from their parents as infants and raised in an environment in which disease X is absent? If they do develop disease X, the case for "nature" is strong.

However, it is not ironclad. Like twin studies, adoption studies have drawbacks:

1. Are the real parents really the real parents? In the case of the mother one can be fairly sure. But what about the father? Many children placed for adoption are illegitimate. If the mother is sure about the identity of the father, she may not reveal it; in some cases she may not be sure.
2. By the time a child is adopted, even if early in life, he has been exposed to environmental influences for many months. He has spent nine months in the uterus, exposed to a torrent of physical and chemical events, and the mother often nurses the child and otherwise cares for him at least for a short period after birth. It seems unlikely that these experiences would increase a susceptibility to alcoholism, but who can be sure? If the woman is a drinker, it is certain that the fetus was exposed to alcohol, and if the mother drank alcohol while nursing, the infant also drank alcohol while nursing. Again, it is unlikely that exposure to alcohol in fetal life or via maternal milk would result many years later in alcoholism, but it is hard to prove that it might not happen.
3. Sexual or marital partners do not choose each other entirely by chance. People tend to marry people of similar education and social backgrounds, for example, and probably also marry people with similar habits. Alcoholic men may be likely to marry women with a proclivity to drink. A genetic susceptibility to alcoholism provided by the father might then be compounded by genetic tendencies provided by the mother.
4. Adoption studies are difficult. Adoption agencies are often reluctant to cooperate, sometimes for good reasons, and often have little information about the biological parents. Also, tracking down people who were adopted twenty or

thirty years ago is not easy, particularly in a mobile society such as the United States. For this reason, adoption studies are rare.

GENETIC MARKERS

Some traits are definitely inherited. In addition to those already mentioned—blood groups, colorblindness—a number of blood proteins and the ability to taste certain substances are among them. Because these traits are unquestionably genetic and the frequency of their occurrence in families and populations has been studied, they are sometimes used as "genetic markers" to study illnesses that *may* be hereditary.

Consider, again, colorblindness. If every colorblind individual in a family is alcoholic, and no noncolorblind individuals are alcoholic, there would be a strong implication that alcoholism, like colorblindness, was inherited. In short, color-association would be a "marker" for alcoholism. Moreover, this coincidence would provide information about how alcoholism was inherited. The gene for colorblindness is located at a particular position on the X chromosome, the one that determines sex. If colorblindness and alcoholism coexist, the "gene" for alcoholism would also seem to be located on the X chromosome, near the colorblindness gene. There is evidence that this is true, but also evidence that it is not (see chapter 5).

Marker studies are of two types: pedigree studies, in which single families are studied, and population studies. In the latter, the occurrence of genetic markers in a population is compared with the occurrence in individuals with particular diseases. The field of population genetics is still primitive, and there is conflicting evidence about the wide variation in the frequency with which markers occur in populations. Nevertheless, if a certain blood group apparently occurred in 30 percent of a given population, but 90 percent of the alcoholics in the population had the blood group, it would suggest a genetic connection. There is some evidence for such a connection, but, again, it is weak.

The main problem with the genetic-marker approach is that it would be a fantastic coincidence if the genes for the illness being studied happened to be located an infinitesimal distance from the genes for the marker being studied. Genetic-marker studies are needle-in-haystack studies, and the needles have been elusive.

Predictors

Markers are predictors. If alcoholics more often have blood group A than nonalcoholics, group A is a predictor of alcoholism. Not all predictors must be genetic, like blood groups. Certain behaviors or psychological tests may predict alcoholism. The important thing is that the behavior or tests must precede the onset of heavy drinking. Differences between alcoholics and nonalcoholics, whether psychological or physical, are difficult to interpret, and may be consequences of drinking, not possible causes of drinking.

One way out of this dilemma is to study children of alcoholics before they start drinking. Based on most studies, at least 20 or 25 percent of the sons of alcoholics eventually develop drinking problems; they are said to be at high risk for alcoholism. At least a dozen institutions in several countries are studying children of alcoholics; the studies are called high-risk studies. The goal is to obtain a wide range of predrinking data, follow the individuals into mid-life, and then identify predrinking data that predict future drinking problems.

So far, none of the studies have gone beyond the predrinking data collection stage. Follow-up studies to identify predictors are planned when the subjects are in their thirties and forties. Meanwhile, in the initial predrinking studies, interesting differences have been found between sons of alcoholics and sons of nonalcoholics. (The subjects are almost always males because men become alcoholic more often than women and thus fewer need be studied to identify a sizable group of alcoholics.) Possible predictors, so far identified, include the following:

1. Sons of alcoholics generate more alpha activity on the electroencephalogram (EEG) after drinking alcohol than do sons of nonalcoholics. Earlier studies found reduced alpha activity in alcoholics, usually interpreted as representing damage from heavy drinking. The finding that adolescent sons of alcoholics who were not heavy drinkers had increased amounts of alpha activity after drinking suggests that a risk factor in alcoholism may be a genetically determined increase in alpha activity from alcohol.

This finding is especially interesting because alpha waves are associated with relaxation and serenity. There are biofeedback gadgets on the market to help people increase their alpha activity through conditioning. Transcendental meditation and Zen Buddhism are supposed to have the same effect. Perhaps sons of alcoholics become alcoholic because alcohol enhances the alpha-related events in the brain associated with serenity and thus is more rewarding to them than to sons of nonalcoholics. It must be stressed that not all sons of alcoholics show increased alpha activity after drinking. If, some day, it turns out that those who showed increased alpha activity were also the ones who became alcoholic, this will be strong supporting evidence for the theory that perhaps alcoholism should be renamed alphaholism, as some wit suggested.

2. Sons of alcoholics do less well on a particular test for abstract or "conceptual" thinking called the categories test. The test involves identifying organizational motifs in groups of symbols. This also is interesting. Alcoholics also do poorly on the categories test. This is usually seen as evidence of brain damage from drinking. Since their nondrinking sons as a group also do poorly on the test, poor performance on the test may turn out to be an antecedent of alcoholism rather than a consequence.

3. One study found that, compared to controls, sons of alcoholics had modestly elevated blood levels of acetaldehyde after drinking alcohol. Acetaldehyde is a breakdown product of alcohol and even after heavy drinking exists in only minute quantities in the blood. This is fortunate because acetaldehyde is very toxic. (The substance Antabuse combined with alcohol

produces an adverse reaction precisely because the combination increases the level of acetaldehyde.) If sons of alcoholics do have modestly elevated levels, this would be a useful predictor of alcoholism. It also may have causal significance. Modest elevation of acetaldehyde levels in animals is associated with increased drinking. Possibly, modest elevation may encourage drinking, whereas higher elevation may deter drinking. (Antabuse produces high elevation.)

4. Sons of alcoholics feel less intoxicated after drinking the same amount of alcohol as sons of nonalcoholics. They are also behaviorally more tolerant of alcohol than low-risk subjects (i.e., they show fewer effects on the same amount). From this and other evidence it appears that tolerance for alcohol (ability to drink a lot) is probably more innate than acquired.

5. Henri Begleiter and associates at Downstate Medical Center of New York University in Brooklyn studied more than 200 sons of alcoholics, ranging in age from six to late adolescence, and compared them to 200 controls. They found that sons of alcoholics responded to clicks or flashes of light with a different EEG wave than did sons of nonalcoholics. Specifically, the difference appeared in the so-called P300 wave that occurs about 300 milliseconds after the click or flash of light. As a group, the sons of alcoholics had a decreased amplitude of the P300 wave after the stimulus. Apparently their brains do not organize a response to the stimulus with the same degree of efficiency as occurs in the brains of control subjects.

The P300 wave reflects a person's ability to identify "relevant" stimuli. This suggests that sons of alcoholics are in some way deficient in this ability. Whatever implications this has for their future drinking habits, cognitive skills, or personality factors is not known. However, the finding is highly consistent across studies. At least six studies have repeated the observation.

Begleiter also found that the association of decreased amplitude of P300 occurred with a high degree of regularity in the subtype of alcoholism described by Bohman and Cloninger (see chapter 5) as Type 2 alcoholism (which this author believes closely resembles familial alcoholism). When only Type

2 sons were studied, the P300 abnormality identified them in 87 percent of the cases, compared with 35 percent of the cases when there was a mixture of Type 1 and Type 2. This has been the only opportunity, so far, to validate the importance of familial (Type 2) versus nonfamilial (Type 1) alcoholism in a large-scale study.

In two other studies, sons of alcoholics were given alcohol before the EEG was recorded. In one study there appeared to be diminished amplitude of the P300, but this was not found in the second study. A third study found that mere expectation of receiving alcohol caused a large reduction of P300 amplitude in the sons of alcoholics. Alcohol expectancy produced no change in P300 amplitude in the sons of nonalcoholic fathers. The importance of set (expectation) was stressed in chapter 1; this may be the first demonstration of a physiological correlation with set, at least in alcohol studies.

In any case, the P300 finding, one of the most consistent observations in alcoholism research, is relevant to the familial versus nonfamilial studies in chapter 3 and adoption studies in chapter 5. It is one of the most interesting findings of the past decade.

Whether these differences between sons of alcoholics and sons of nonalcoholics turn out to be predictors of future alcoholism will only be known after follow-up studies are performed and correlations made between data collected before the subjects had experience with alcohol and subsequent drinking behavior. It will probably be ten years before such information is available.

Meanwhile, current alcoholism literature provides little help in identifying people in adolescence who will later have drinking problems. There have been only four longitudinal studies of sufficient size and methodologic rigor to offer the promise to provide useful information, and they have provided little indeed. What they mainly find is an absence of predictors in adolescence.

The first of these studies, conducted by McCord and McCord, reported that adolescents who later had drinking problems were more assertive and impulsive and had fewer fears than adolescents who did not have problems later. A study by Jones had similar results. As mentioned in chapter 2, Vaillant actually found that those adolescents who went on to become alcoholic seemed as healthy as the subjects who did not develop drinking problems. The future alcoholics not only had many features identified with "good" mental health but also came from stable families. They were self-confident, gregarious persons whose dependency needs (if present at all) were marvelously concealed. In longitudinal studies of whites and blacks in St. Louis, Robins and coworkers found that school failure and truancy predicted future alcoholism. However, the groups were atypical, with whites identified from court-referred treatment cases and blacks from a large-city ghetto.

Some retrospective studies, but not all, tend to support the idea that antisocial behavior in childhood may predict future alcoholism. In these studies, alcoholics more often give a history of hyperactivity and/or conduct disorder in childhood and early adolescence than do nonalcoholics. Most hyperactive children, however, do not become alcoholic.

There are also "softer" predictors. One involves choice of occupation. Alcoholism is unevenly distributed among occupations. More bartenders and journalists are alcoholic than ministers and mail carriers. Perhaps the highest rate of alcoholism reported by any defined group applies to Americans who have won the Nobel Prize for literature. Of the seven Americans who won the prize, four were alcoholic and a fifth was a heavy drinker.

Ethnic background is another predictor. For example, based on a large number of studies, Irish immigrants have a relatively high rate of alcoholism, Jews a low rate.

A strong predictor, of course, is gender. As discussed, men are three to five times more likely to become alcoholic than women.

The strongest predictor of future alcoholism, however, is a family history of alcoholism. The next two chapters address the problem of whether transmission of alcoholism in families is based mainly on inherited or environmental factors.

CHAPTER FIVE

The Case for Nature

There is no escape from the conclusion that nature prevails enormously over nurture when the differences in nurture do not exceed what is commonly to be found among persons of the same rank of society in the same country. My only fear is that my evidence seems to prove too much, and may be discredited on that account, as it seems contrary to all expectation that nurture should go for so little. But experience is often fallacious in ascribing great effects to trying circumstances. Many a person has amused himself by throwing bits of stick into a brook and watching their progress; how they are arrested, first by one chance obstacle, then by another, and again, how their onward course is facilitated by a combination of circumstances. He might ascribe much importance to each of these events, and think how largely the destiny of the stick has been governed by a series of trifling accidents. Nevertheless, all the sticks succeed in passing down the current, and they travel, in the long run, at nearly the same rate. So it is with life, in respect to the several accidents which seem to have had a great effect upon our careers. The one element, which varies in different individuals, but is constant for each of them is the natural tendency; it corresponds to the current in the stream, and inevitably asserts itself.

—Francis Galton

Chapter 4 discussed various strategies for separating nature from nurture—adoption studies, twin studies, and genetic marker studies. Each strategy has been applied to alcoholism. In addition, an animal model for alcoholism has been produced that allows further insights. I'll examine each of these strategies in turn as they apply to the role of heredity—"nature"—in alcoholism.

ADOPTION STUDIES

There have been four adoption studies. Because of my personal involvement in one of them—a study conducted in Denmark in the 1970s and 1980s—I will describe this work in detail. The study lasted six years and went through four phases:

Phase 1. Sons of alcoholics were compared to sons of nonalcoholics where both groups had been adopted in infancy.

Phase 2. Sons of alcoholics adopted in infancy were compared to male siblings raised by the alcoholic parents.

Phase 3. Daughters of alcoholics were compared to daughters of nonalcoholics where both groups had been adopted in infancy.

Phase 4. Daughters of alcoholics adopted in infancy were compared to female siblings raised by the alcoholic parents.

The goals and methodology for each phase were identical and are presented in the section on Phase 1.

The Danish studies: phase 1

Adoption studies pose many problems. For example, obtaining access to adoption agency records may be difficult. Most agencies have little information about the parents of children who are placed for adoption. Also, among highly mobile people such as Americans, locating subjects is difficult.

For these reasons our collaborative adoption study was done in Denmark, a country where few of these difficulties exist. There is little population movement in or out of the country, and centralized national registries are available for scientific purposes.

Originally, the study had two goals: (1) to determine whether men raised apart from their biological parents were more likely to have drinking problems or other psychiatric difficulties if one of their biological parents was alcoholic than if there was no recorded alcoholism among their biological parents; (2) to determine whether sons of alcoholics raised by their alcoholic parents were more likely to develop alcoholism than were their brothers who had been adopted in early infancy and raised by foster parents who were not alcoholic. The study was later expanded to include daughters of alcoholics.

The subjects for the study were obtained from a pool of some 5,000 nonfamily adoption cases—all that had taken place in the city and county of Copenhagen from 1924 through 1947. This pool of adoption cases was originally established as part of a family study of schizophrenia. As part of the same study, hospital records were screened to identify those adoptees and their biological and adoptive parents who had been psychiatrically hospitalized. Selected for our initial study were adopted men with a biological parent who had been hospitalized for alcoholism (hereafter referred to as "probands"). All had been separated from their biological parents in early infancy, had been adopted by nonrelatives, and had no known subsequent contact with their biological relatives. Only male adoptees were studied, since more men become alcoholic than women. Eighty-five percent of the alcoholic parents were fathers.

Another group of adoptees, the "control" group, also was studied. It differed from the proband group in one regard: The men had not had a biological parent hospitalized for alcoholism. The two groups had the same age composition and the men in both groups had been adopted at the same age by nonrelatives.

A Danish psychiatrist interviewed the two groups without knowing whether they were probands or controls. The interviews were in Danish but recorded in English. They lasted two to four hours and had a detailed format for eliciting information about social and economic factors, the adoptive parents, psychiatric illness in the adoptees, drinking practices and problems, and a wide range of other life experiences. The study was not identified as an adoption study, and subjects rarely volunteered information about their biological parents; no subject mentioned that he had a biological parent with a drinking problem. Law-enforcement records also were reviewed for information about the adoptees and their biological parents.

The interviews and records were sent to Washington University in St. Louis, where computer analysis was performed, again without knowledge of the subjects' category. The study was blind from beginning to end.

A total of 133 subjects were located and cooperated in the study. They consisted of 55 adoptees with an alcoholic biological parent (the probands) and 78 adoptees (controls) with no known alcoholism in their biological parents.

Table 1 compares the proband and control groups. The mean age in both groups was thirty, with a range in age of twenty-three to forty-five years at the time of interview. The socioeconomic status and educational experience of the men in both groups were quite similar. Divorce was the only factor in Table 1 that significantly distinguished the two groups. The probands had a divorce rate three times greater than that of the control group.*

Table 2 presents information about the adoptive parents. The home was intact when the subject lived with both adop-

*When the term *significant* is used, it means that the differences between the two groups were unlikely to have happened by chance, as determined by statistical tests.

Table 1. CHARACTERISTICS OF ADOPTEES (%)

	Probands* (N = 55)	Controls† (N = 78)
Age 20–24	9‡	10
25–29	51	53
30–34	14	19
35–39	14	10
>40	11	8
Marital history		
Ever married (never divorced)	62	74
Ever divorced	27	9
Never married	11	17
Academic education		
<9 yr	67	69
9–11 yr	24	23
12–15 yr	7	8
>15 yr	2	0
Socioeconomic class		
Low	53	46
High	47	54
Served in military	67	72

*Refers in this and other tables to adoptees of alcoholic parentage.
†Refers to adoptees of nonalcoholic parentage.
‡Figures rounded off to nearest whole number for ease of perusal. Sums may therefore not equal 100 percent.

tive parents from adoption until at least age sixteen. Estimates of psychiatric illness in the adoptive parents were based on information provided by their adopted children. The parents were considered depressed, alcoholic, or suffering from some form of psychiatric illness if the condition led them to seek treatment.

In general, the two groups of adoptive parents were similar. The adoptive parents of the probands tended to have some-

Table 2. FOSTER-HOME EXPERIENCE (%)

Adoptive-Home Characteristics	Probands (N = 55)	Controls (N = 78)
Home intact	87	86
Parental economic status		
Below average	15	19
Average	58	55
Above average	27	26
Parental psychopathology, any	42	50
Father depressed*		
Possible	2	8
Definite	7	4
Father alcoholic		
Possible	5	10
Definite	7	12
Father antisocial		
Possible	0	3
Definite	0	0
Father, any psychopathology	24	36
Mother depressed		
Possible	4	5
Definite	16	9
Mother, any psychopathology	29	28

*Diagnoses are based on the adoptees' observations and do not represent firsthand psychiatric observations.

what more depression than did the adoptive parents of the controls, but the differences were small. The alcoholism rate actually was higher in the adoptive parents of the control group, another presumably chance occurrence.

Table 3 compares psychiatric illnesses in the adoptees. Again, the differences were small. Total illness rates, exclusive of alcoholism, were almost identical.

Table 4 concerns the subjects' drinking behavior. Almost without exception, the proband group had more drinking problems than did the control group. With regard to five variables, the differences between the two groups were statistically significant.

"Hallucinations" in Table 4 refers to auditory or visual per-

Table 3. PSYCHIATRIC DISORDERS IN ADOPTEES (%)

	Probands (N = 55)	Controls (N = 78)
Anxiety neurosis		
Possible	16	18
Definite	2	10
Depression		
Possible	13	17
Definite	2	3
Criminality		
Possible	0	3
Definite	5	3
Drug abuse		
Possible	4	1
Definite	5	1
Personality disturbances*	51	53
Psychopathology, any (excluding alcoholism)	56	53

*Refers to a variety of character diagnoses commonly used in Denmark; for example, antiaggressive personality, compulsive personality, sensitive-insecure personality. Most of these are not synonymous with "psychiatric illness" as the term is commonly used.

ceptual distortions associated with withdrawal from alcohol. "Lost control" refers to the experience of wanting to stop drinking on a particular drinking occasion but being unable to do so. "Morning drinking" refers to repeated morning drinking, not just one or two before-breakfast drinking experiences. "Rum fits" are epileptic seizures that occur only after withdrawal from alcohol.

The men in the proband group had an average of 2.05 alcohol-related problems. Those in the control group had an average of 1.23 alcohol-related problems. The difference was significant.

Table 4 also compares the two groups with regard to how many had been treated for alcoholism. Nine probands had been treated, compared with one control. Seven probands but no controls had been hospitalized.

The last items in Table 4 present a composite picture of

Table 4. DRINKING PROBLEMS AND PATTERNS IN TWO ADOPTIVE GROUPS (%)

	Probands ($N = 55$)	Controls ($N = 78$)
Hallucinations*	6	0
Lost control*	35	17
Amnesia	53	41
Tremor	24	22
Morning drinking*	29	11
Delirium tremens	6	1
Rum fits	2	0
Social disapproval	6	8
Marital trouble	18	9
Job trouble	7	3
Drunken-driving arrests	7	4
Police trouble, other	15	8
Treated for drinking, ever*	16	1
Hospitalized for drinking	11	0
Drinking pattern		
Moderate drinker	51	45
Heavy drinker, ever	22	36
Problem drinker, ever	9	14
Alcoholic, ever*	18	5

*These differences between the groups are statistically significant, meaning they are unlikely to have occurred by chance.

drinking patterns based on quantity and frequency of drinking as well as problems associated with drinking. Criteria for the four drinking categories are given in Table 5. Using these criteria, ten of the fifty-five probands were "alcoholic" as compared with four controls. In other words, the probands had nearly four times the alcoholism rate of the control group. Heavy and problem drinking occurred somewhat more frequently in the control group, but *only alcoholism significantly distinguished the two groups.*

In short, the study indicated that children of alcoholics are more likely to have alcohol problems than are children of nonalcoholics, despite being separated from their alcoholic

Table 5. CRITERIA FOR DRINKING CATEGORIES

Category	Criteria
Moderate drinker	—Neither a teetotaler nor heavy drinker.
Heavy drinker	—For at least one year drank daily and had six or more drinks at least two or three times a month; or drank six or more drinks at least once a week for more than one year, but reported no problems.
Problem drinker	—(a) Meets criteria for heavy drinker. (b) Had problems from drinking but insufficient in number to meet alcoholism criteria.
Alcoholic	—(a) Meets criteria for heavy drinker. (b) Must have had alcohol problems in at least three of the following four groups: GROUP 1: Social disapproval of drinking by friends, parents Marital problems from drinking GROUP 2: Job trouble from drinking Traffic arrests from drinking Other police trouble from drinking GROUP 3: Frequent blackouts Tremor Withdrawal hallucinations Withdrawal convulsions Delirium tremens GROUP 4: Loss of control Morning drinking

parents in early life. All individuals in the study were separated from their biological parents in early infancy and raised by nonrelatives without subsequent contact with their natural parents. Nevertheless, by their late twenties or earlier the offspring of alcoholics had nearly twice the number of alcohol problems and four times the rate of alcoholism as the children whose parents had no record of hospitalization for alcoholism.

Alcoholism is a term that lends itself to so many definitions that whichever one is chosen can be challenged by those who disagree with it. It should be emphasized, therefore, that use of the term in this study was based on arbitrary criteria; a certain minimum number of problems were required before the diagnosis was made, and no single problem was considered indispensable to the diagnosis.

Perhaps the most objective measure of the greater number of alcohol problems in the proband group was the extent of psychiatric treatment in the group. Eight of the fifty-five probands had been psychiatrically hospitalized, as compared with two of the seventy-eight controls. Moreover, seven of the eight probands hospitalized were diagnosed as alcoholic by the criteria in Table 5. None of the four controls with a diagnosis of alcoholism had been hospitalized. Nine of the ten alcoholic probands had received psychiatric treatment, compared with one of the four alcoholic controls.

Apart from alcoholism, the two groups were virtually indistinguishable with regard to other types of psychiatric illness and life experiences, with one exception: divorce. The probands were three times more likely to be divorced than the controls. Drinking did not appear to be related to the higher divorce rate. Moderate drinkers in the proband group were as likely to be divorced as heavy drinkers and alcoholics. Divorce and alcoholism have often been associated, with divorce usually attributed to the disruptive effects of alcoholism. The explanation may be more complicated.

Concerning both alcoholism and divorce, it is important to note that most of the subjects in the study have by no means traversed the age of risk for either of these problems. More than 60 percent of the subjects were still in their twenties at the time of interview. Studies indicate that the period of risk for alcoholism is roughly age twenty through forty-five. Therefore, the alcoholism and divorce rates may continue to increase over the next two decades, although whether the difference between the proband and control groups will continue to widen cannot be predicted.

The finding that, apart from divorce, *only* alcoholism significantly distinguished the two groups suggests there may be a specificity in the transmission of the disorder heretofore underestimated. The rates of diagnosable depression, anxiety neurosis, criminality, and drug addiction were fairly low in both groups, and neither group had a significantly higher rate of any one of these conditions than did the other. Also, it is interesting that heavy and even problem drinking, as defined in this study, failed to distinguish the two groups. If anything, there was somewhat more heavy and problem drinking in the control group than in the proband group. *This suggests that severe forms of alcohol abuse may have a genetic predisposition but that heavy drinking itself, even when responsible for occasional problems, reflects predominantly nongenetic factors.*

It should be emphasized that "genetic predisposition" remains more probable than proven, and certainly may not apply to all alcoholics. Even the possibility of environmental influence cannot be entirely ruled out. The adoptees as a rule spent the first few weeks of life in the care of their biological mothers. Wives of alcoholics may well differ from wives of nonalcoholics, and conceivably the mothers of the probands, when not alcoholic themselves, had other forms of psychiatric illness leading to neglect or other kinds of harmful influence on their children. This explanation of the study's findings, however, does seem farfetched.

Another nongenetic factor that may have biased the results was selectivity in the process of adoption. Possibly those infants that had a known alcoholic parent at the time of adoption may have been matched with less "desirable" parents than children of nonalcoholic parents. Adoption practices in Denmark twenty to fifty years ago are difficult to assess. From an adoption study of criminality also conducted in Denmark, it appears that a bias did occur in the adoption process, but only at the extreme ends of the social spectrum. In other words, upper-class adoptive parents tended to receive upper-class biological children, and lower-class adoptive parents received lower-class adoptive children. In the very large middle-class group, however, there was no evidence of bias. Moreover, since the adoptive parents of the two groups did not differ with respect to educational or economic status, it would appear that a selective bias in adoption was minimal at most.

The Danish studies: phase 2

In some instances, the adopted sons of the alcoholics had biological siblings raised by the alcoholic parent. In the second phase of the Danish study we compared the drinking problems and psychiatric disorders of children of alcoholics raised by their biological parents with those of their brothers who had been adopted early in life and raised in a different environment.

Twenty adopted sons of alcoholics had brothers who had been raised by their alcoholic parent and who were available for interview. Thirty of these brothers were interviewed by a Danish psychiatrist, using the interviewing method employed in the first phase of the study. To avoid bias, interviews also were conducted with fifty nonadopted control subjects selected from census records. All interviews were blind, the interviewer having no knowledge whether he was dealing with sons of alcoholics or with controls.

As a group, the nonadopted sons of alcoholics differed from their adopted brothers in the following ways:

1. They were older by an average of three years.
2. They were less likely to have been married.
3. As adults, they belonged to a lower social class, and their parents-of-upbringing also were of a lower social class.
4. Their parents-of-upbringing had higher rates of alcoholism and were more often "antisocial" than the adoptive parents. The former was predictable, since it was known from hospital records that *all* of the parents of the nonadopted children had received a diagnosis of alcoholism at some time.

Table 6 shows problems associated with drinking. In general, the nonadopted sons had no more problems than the adopted sons. The last four items represent a composite picture of drinking patterns, based on the criteria given in Table 5. Using these criteria, five men in each group were considered alcoholic, resulting in a higher rate of alcoholism in the adoptees, but not a statistically significant difference. There was some evidence that as the severity of the parents' alcoholism

Table 6. COMPARISON OF DRINKING PROBLEMS AND PATTERNS (%)

	Adopted Sons ($N = 20$)	Nonadopted Sons ($N = 30$)
Hallucinations	10	10
Lost control	25	10
Amnesia	40	13
Tremor	10	14
Morning drinking	30	21
Delirium tremens	10	3
Rum fits	0	0
Social disapproval	10	20
Marital trouble	0	7
Job trouble	10	10
Drunken-driving arrests	0	3
Police trouble, other	5	10
Treated for drinking, ever	15	13
Hospitalized for drinking	10	10
Drinking pattern		
Moderate drinker	60	77
Heavy drinker, ever	15	3
Problem drinker, ever	0	3
Alcoholic, ever	25	17

It is interesting how little environment appeared to contribute to the development of alcoholism among the sons of alcoholics in this sample. Our findings tend to contradict the oft-repeated assertion that alcoholism results from the interaction of multiple causes—social, psychological, biological. This may be true of milder forms of alcoholism, but conceivably severe alcoholism could be relatively uninfluenced by environment, given free access to alcohol. The "father's sins" may be visited on the sons even in the father's absence. (The possibility that severe forms of alcoholism are influenced by heredity, but less severe forms are not, is a recurring theme in this chapter.)

increased (as measured by number of hospitalizations), the tendency toward alcohol problems in the sons increased.

The main finding of the study was that sons of alcoholics were no more likely to become alcoholic if they were reared by their alcoholic parent than if they were separated from their alcoholic parent soon after birth and reared by nonrelatives. This was true despite the fact that, as a group, the nonadopted sons were older than their adopted brothers and, therefore, further advanced into the age of risk for alcoholism.

Compared with the adoptees, there was also evidence that the children exposed to their alcoholic parent were raised in environments associated in other studies with increased risk of alcoholism. Their parents-of-upbringing were of a relatively low social class, and the subjects themselves were of lower socioeconomic status than their adopted brothers (it has often been found that there is an increased risk of alcoholism among the lower classes). Furthermore, the nonadopted children appeared to have had a more disruptive childhood than their adopted brothers. More of them received welfare help as children; more did poorly in school; more had a parental history of psychiatric illness; and all these factors are associated with an increased risk of alcoholism.

The Danish studies: phase 3

Daughters of alcoholics were investigated in the third phase of the adoption study. The study sample consisted of forty-nine daughters of alcoholics, raised by adoptive parents, and, as controls, forty-eight daughters of presumed nonalcoholics also raised by adoptive parents. Subjects and controls both had a mean age of thirty-five years. As was the case in former studies, interviews were conducted blindly to eliminate interviewer bias. The major findings from this phase were as follows:

1. Of both the sample and the control group, 4 percent was either alcoholic or had a serious drinking problem. The sample was too small to draw definite conclusions, but as the estimated prevalence of alcoholism among Danish women is about 0.1 to 1.0 percent, the data suggest there may be an increased prevalence of alcoholism in the two groups. Nothing was known about the controls other than that none had a biological parent with a hospital diagnosis of alcoholism. Explanation of the appearance of alcoholism among controls

was not readily apparent; however, it was interesting that both groups of women had adoptive parents described as alcoholic, suggesting that environmental exposure to alcoholism may contribute to alcoholism in women (although it did not appear to in men, as demonstrated by data from Phase 2 of this study). It also should be noted that although controls were presumed to have nonalcoholic biological parents, this does not preclude their having parents with drinking problems, because the only index for the controls was the lack of a hospital diagnosis of alcoholism in their parents.

2. More than 90 percent of the women in both groups were abstainers or very light drinkers. This contrasted with the case of the Danish male adoptees of whom about 40 percent were heavy drinkers.

3. Family history studies have suggested that alcoholics often have male relatives who are alcoholic or sociopathic and female relatives who are depressed. In this study, however, there was a low rate of depression in both groups, with no more depression in the daughters of the alcoholics than in the control women.

From this study, it appears that alcoholism in women may have a partial genetic basis, but the sample size precluded any definitive conclusion. Since the great majority of the women drank very little, it is possible that social factors discouraging heavy drinking may suppress a genetic tendency, if indeed one exists. There was no evidence of a genetic predisposition to depression in the adopted daughters of alcoholics. It should be noted that some evidence indicates that women develop alcoholism at a later age than do men, and undoubtedly some of these thirty-five-year-old women had not yet traversed the age group of risk for either alcoholism or depression.

The Danish studies: phase 4

In the fourth phase, drinking problems and other psychopathology in daughters of alcoholics raised by their alcoholic biological parents was compared with that seen in daughters of alcoholics raised by adoptive parents. Forty-nine adopted and eighty-one nonadopted women were interviewed. The average age of the adopted daughters was thirty-seven, compared to thirty-two in the nonadopted daughters, a significant difference. More of the adopted daughters had been divorced and more of the nonadopted daughters had never married—both factors possibly related to the age difference. The nonadopted women tended to come from a lower socioeconomic class. Adoptees more often lived with an "intact" family (i.e., the same mother and father until at least age sixteen).

More of the adoptees had received psychiatric treatment. Based on other studies, adoptees generally receive more psychiatric treatment than nonadoptees for reasons not necessarily related to mental illness. Sensitivity of the foster parents to emotional problems in their adopted children may partly account for this.

Applying the criteria described in the section on Phase 1, 2 percent of the adopted daughters and 3 percent of the nonadopted daughters were alcoholic. Among the adopted daughters, 2 percent were problem drinkers. Like those identified as alcoholic, the problem drinkers also had been hospitalized for drinking and thus probably would be considered alcoholic. Estimates of alcoholism among Danish women range from 0.1 percent to 1 percent. Thus alcoholism rates of 3 percent to 4 percent in the daughters of alcoholics exceed the expected rate. Although this represents a fourfold increase over the expected rate, it is still far less than the percentage reported for the sons of alcoholics.

This study of daughters of alcoholics produced other findings differing in several respects from those found in the study of sons of alcoholics.

A much higher percentage of the daughters were light or moderate drinkers. More than 90 percent of the daughters drank very little; almost half of the sons were heavy drinkers. Correspondingly, the alcoholism rate among the sons was higher. About 18 percent of the sons (whether adopted or nonadopted) were alcoholic compared to an estimated prevalence rate of alcoholism in Denmark of about 5 percent in the male population.

The lower rate of alcoholism among women may be related to the finding that far more women were light or moderate drinkers than were the men. Of those women who did drink immoderately, most developed problems requiring treatment. This was not true of Danish men; nearly half were heavy drinkers but not alcoholic.

Cultural or biological factors (or both) may discourage many women from having more than an occasional drink. Generally, in Western countries drunkenness is more condemned in women than in men; indeed, until recently, any drinking by women in some regions was considered immoral. In addition, as a group, women may be physiologically more susceptible to adverse effects from alcohol, such as headache or nausea, and therefore be somewhat "protected" from overindulgence.

Family studies of alcoholism emphasize the high prevalence of alcoholism among male relatives of alcoholics. In some families, at least 25 percent of close male relatives are alcoholic. However, these studies also show a higher than expected frequency of alcoholism among female relatives. It is possible that daughters of alcoholics are as genetically susceptible to alcoholism as sons, but since fewer drink, fewer become alcoholic.

Another difference between the studies of sons and daughters was that the sons differed from controls in reporting increased rates of childhood hyperactivity, truancy, and school phobia. This did not occur with the daughters. On the other hand, daughters were more often friendless as children and shy as adults, suggesting a relationship between certain personality traits and a family history of alcoholism.

Studies of female alcoholism often indicate that women begin drinking heavily at a later age than men. The Danish data support this observation. Almost all of the men who were problem drinkers or alcoholic began drinking heavily in their teens or early twenties, whereas almost all of the alcoholic women began drinking heavily after the age of thirty.

Also, it has been reported that women become alcoholic at a later age. Most of the women in this study were in their early or mid-thirties. A follow-up study might show higher rates of alcoholism as the women became older.

The studies of sons and daughters had at least one feature in common: Having a biological parent who was alcoholic increased the risk of the child's becoming alcoholic, but, if the child was raised by foster parents, having a biological father who was alcoholic did *not* increase the possibility of other psychiatric disorders occurring. In other words, the susceptibility to alcoholism was specific for alcoholism—not on a continuum with heavy drinking, and not part of a spectrum with other illnesses.

However, in the case of nonadopted daughters, there was an overlap of alcoholism and depression. Daughters of alcoholics raised by their alcoholic parents had a history of depression more often than occurred in the age-controlled general population (27 percent vs. 7 percent).

This is an interesting finding in view of a hypothesis proposed by George Winokur of the University of Iowa that alcoholism and depression are related in families. On the basis of his own studies and those of several other investigators, he proposed that the "typical" family of an alcoholic consisted of alcoholism and sociopathy among male relatives and depression (particularly early-onset depression) among female relatives.

Our study supported the idea of a connection between alcoholism and depression, but the overlap only occurred in daughters raised by their alcoholic parents. As with many known genetic disorders, alcoholism in women may require environmental precipitants, and therefore our finding does not rule out the possibility of a genetic predisposition to depression among daughters of alcoholics. Nor, however, does it support the hypothesis, since genetic and environmental influences are usually inseparable in persons raised by their own biological parents.

The sons and daughters studied shared another common feature. Both the adopted sons and adopted daughters had divorce rates significantly higher than did their respective adoptee controls. This was not true of nonadopted sons and daughters. This may be due to happenstance, but the fact that it occurred in both studies should at least generate some speculation about a possible genetic–environmental interaction between divorce and alcoholism. The divorced adoptees, by the way, were not the alcoholics or heavy drinkers, so drinking itself did not appear to be the cause of divorce.

The Swedish studies

In 1978, Michael Bohman, a Swedish pediatrician, obtained access to so-called temperance records maintained throughout Sweden. (Whenever a Swedish citizen was arrested for an alcohol-related crime, such as drunken driving or disorderly conduct, or received treatment for a drinking problem, he or she received a citation in a temperance record.) Bohman also obtained records for all persons born out of wedlock in Stockholm from 1930 through 1949 and placed for adoption. Most of the children were separated from their biological relatives in the first few months of life. The average age at final placement in the adoptive home was eight months.

Like Denmark, Sweden is an ideal site for an adoption study. Extensive social and medical records about the adoptees and their parents (both biological and adoptee) were obtained from public sources by Bohman and his associates. Alcohol is readily available in Sweden, but the Temperance Board in each community is legally responsible for maintaining sobriety. The board imposes fines for intemperance, supervises the treatment of alcoholics, and keeps records, which were available to Bohman. Other records about diagnoses and treatment are kept by agencies of the National Health Insurance. Arrest records were also available.

Bohman found that biological parents with multiple entries in the Temperance Board records were particularly likely to have adopted-out sons with multiple entries. From other studies he knew that individuals with multiple entries closely fit the clinical picture of alcoholism. As in the Danish studies, alcoholic fathers seemed to beget alcoholic sons, even when the sons were raised by nonalcoholic adoptive parents.

On the other hand, criminality did not predict alcoholism in the adopted-out children, nor did alcoholism in the biological parents predict criminality in the adoptees. Thus, in a study based entirely on records, Bohman came up with substantially the same findings as the Danish study. Alcoholism was transmitted from parents to sons even when the sons were raised apart from the parents. The correlation was much weaker with regard to women adoptees but, again, like the Danish study, there were few alcoholic women adoptees in the study.

In the 1980s, C. Robert Cloninger, a professor at Washington University in St. Louis, began a detailed analysis of the Swedish data using highly sophisticated statistical techniques. He was particularly interested in addressing the following questions:

1. What characteristics of the biological parents influence the risk of alcohol abuse in the adoptees?
2. What characteristics of the adoptive parents influence the risk of alcohol abuse in the adoptees?
3. How do genetic and environmental factors interact in the development of alcohol abuse?
4. Is the genetic predisposition to alcoholism expressed in other psychopathological ways, depending on the environment experience and sex of the individual?

To answer these questions, the investigators selected 862 men and 913 women adopted before the age of three by nonrelatives. The subjects were subdivided according to both congenital background and postnatal home environment. As before, extensive data about alcohol abuse, legal records, occupational status, and medical and social history were obtained for all adoptees and parents from official sources.

Each group was analyzed to determine how adoptees with particular types of congenital background react to different types of adoptive placement. Every possible combination of genetic background and environment was examined. The adoptees were divided into four subgroups according to degree of alcoholism (none, mild, moderate, or severe), based on the number of Temperance Board entries. To determine the role of heredity, several characteristics of the biological parents were examined to identify the ones that were most associated with a particular degree of alcoholism in the adoptees. To understand the role of postnatal environment, a similar analysis of adoptive parents was made to identify the most significant postnatal factors associated with particular degrees of alcoholism.

These studies identified two broad categories of alcoholic families: those in which only men develop alcoholism and those in which both men and women become alcoholic. There was more alcohol abuse in adopted-out sons of alcoholic biological fathers than in the sons of nonalcoholic parents of either sex (22.8 percent of the sons of alcoholic biological fathers were alcoholic, compared to 14.7 percent among sons who did not have an alcoholic biological parent). Also, twice as many of the adopted-out sons whose biological mothers were alcoholic were alcohol abusers than the sons of nonalcoholic parents of either sex (28.1 percent of the sons of alcoholic biological mothers were alcohol abusers, compared to 14.7 percent of sons who did not have an alcoholic biological parent).

The effect of parental alcoholism on adopted women was more complex. The frequency of alcohol abuse was more than three times higher in the adopted-out daughters of alcoholic mothers than in the daughters of nonalcoholic parents (10.8 percent of the daughters of alcoholic biological mothers were alcohol abusers, compared to 2.8 percent of daughters who did not have an alcoholic biological parent). Alcohol abuse in the adoptive parents was not associated with alcohol abuse in the adoptees. Further analysis of the data led to the conclusion that two kinds of genetic predisposition to alcoholism exist—milieu-limited and male-limited. They differ as follows:

Milieu-limited alcoholism is the more common type of genetically influenced alcoholism. It occurs in both men and women (accounting for most alcoholics among both males and females). It is called milieu-limited because its occurrence and severity in susceptible offspring are influenced by the postnatal environment. Thus milieu-limited alcohol abuse requires both a genetic predisposition and environmental provocation. Analysis of the data showed that if only one of these factors was present, the risk of alcohol abuse was lower than in the general population. If both were present, however, the risk was doubled and the severity was determined by the degree of postnatal provocation. (What is meant by "postnatal provocation" will be discussed later.)

Milieu-limited alcohol abuse is usually not severe, typically goes untreated, and is associated with mild, untreated, adult-onset alcohol abuse in either biological parent. The biological parents also show few if any encounters with the legal system. The controlling effect of environment on this form of alcoholism is demonstrated by the finding that a more severe milieu-limited alcoholism requiring hospital care may occur in susceptible sons growing up in particular environments, especially those associated with low social status or unskilled occupation of the adoptive father.

Male-limited alcoholism is a severe type of alcoholism found only in men. It is less prevalent (affecting about 25 percent of male alcoholics in the general population) than milieu-limited alcoholism and appears to be unaffected by the environment. In families with male-limited susceptibility, alcohol abuse was nine times greater in the adopted sons regardless of their postnatal environment. Male-limited susceptibility is associated with severe alcoholism requiring extensive treatment in the biological father, but it is not associated with alcohol abuse in the biological mother. The biological father also tends to have a record of serious law-breaking. Male-limited alcoholism frequently develops in adolescence, is often accompanied by serious encounters with the law, and is associated with episodes of extensive treatment. Criminal behavior, however, appeared to be a consequence of drinking rather than the cause. Postnatal environment has no influence on the risk of developing male-limited alcoholism, but it may influence its severity; adoptees with this form of alcoholism generally were not as severely alcoholic as their fathers, possibly reflecting the advantages of their having been brought up in better homes or other environmental factors.

The male-limited type of susceptibility in a family does not increase the risk of alcohol abuse in the daughters. However, the Swedish adoption studies revealed that it greatly increases the chances of a woman's developing a condition resembling somatization disorder, characterized by frequent complaints of pain or discomfort, without any physiological basis, in various parts of the body.

The investigators caution against premature conclusions. The two types of alcoholism require further study to demonstrate their general validity. The information is based on records, which may not tell the whole story. Judgments about the severity of alcohol abuse derive from temperance records, with one entry denoting mild abuse and four or more entries denoting severe abuse. Entries are a mixture of arrests for a wide range of crimes, ranging from shoplifting to murder, believed by the authorities to be alcohol-related. Alcohol-related offenses are more often committed by men than women, which may explain the small number of women alcoholics in the study. (Of 913 Swedish women adoptees, only 29 had a biological mother, but not a father, who abused alcohol. Ten percent of the daughters were abusers. This amounts to three daughters, a small number to base a conclusion upon.) However, Bohman has assured the author that alcoholism among Swedish women is very low even in population studies.

The authors use the terms *environment* and *postnatal provocation* to refer to social status and occupation. Environment obviously comprises many more variables than social class and occupation. Also, many severely alcoholic individuals, particularly women, may never be cited in the temperance records simply because they do not seek treatment or "get in trouble" because of their drinking. A typical milieu-limited alcoholic would be a son who had a biological parent with mild alcoholism (one entry in the temperance record) but an adoptive parent of low social class and unskilled occupation. A combination of mild alcoholism in the biological family and postnatal provocation in the adoptive family might be sufficient to produce severe alcoholism in the son where it might not occur otherwise.

Regardless of these caveats, the Cloninger–Bohman team produced studies of great interest in the field of alcoholism. Nobody says that genes tell the whole story. Environment, working in subtle and mysterious ways, shapes and modifies all behavior to a greater or lesser extent.

A recent study of combat veterans demonstrates this vividly. In a large sample of Vietnam War veterans, two thirds of combat veterans developed alcohol problems, while two thirds of noncombat veterans did not. Conceivably individuals with characteristics making them more prone to alcohol problems are more likely to be assigned to combat service, but the case can equally be made that many of the veterans with combat-associated alcoholism might have inherited the milieu-limited type of genetic predisposition and needed only a stressful environment to become alcoholic. This area of research has great potential.

In later papers, Bohman and Cloninger relabeled milieu-limited alcoholism as Type 1 alcoholism and male-limited alcoholism as Type 2. By whatever name, they appear to have essentially the same features. As Bohman pointed out, these features resemble those described for familial and nonfamilial alcoholism outlined in chapter 3.

Type 1 is a mild form of alcoholism with an onset in the mid-twenties or later and rarely requiring treatment. Type 2 begins in adolescence or early twenties and is usually severe, often requiring treatment. Type 2 alcoholics more often have parents with severe alcoholism, sometimes more severe than develops in adopted children. In short, Type 2 also could be called familial alcoholism. In the Bohman–Cloninger studies, Type 2 is restricted to men, but this may reflect the fact that information about drinking was obtained from temperance records which, as mentioned, overreport alcohol problems by men and underreport those by women.

Recently, the Bohman group reported that Type 2 alcoholics could be identified by a biological marker. The marker, the enzyme monoamine oxidase (MAO), is produced by blood platelets and is important in the degradation of neurotransmitters associated with emotion in the brain. In a study of thirty-six subjects, Type 2 alcoholics were found to have platelets with lower MAO activity than Type 1 alcoholics. The study was too small to draw conclusions, but it illustrates how the Type 1–Type 2 distinction (by whatever name) may be explored by further study.

The Iowa studies

Everyone agrees that doing adoption studies in the United States is difficult. People are always moving; women tend to change their names. There are no national registries (except Social Security, which doesn't help much). Adoption agencies often do not keep very good records, and even these are often not available. Interviewing people may involve expensive travel from coast to coast. Many people in America resent being interviewed about their personal life.

Despite these obstacles, two adoption studies have been carried out in the United States. The first, conducted by Anne Roe of Yale University in the 1940s, was a small study described in the next section. A more ambitious series of studies has been conducted by Remi Cadoret and associates at the University of Iowa. For the past ten years the Cadoret group has studied adoptees in Iowa. It has been slow going, but the number of subjects has gradually increased over time and there has been a remarkable consistency in the findings reported in a series of papers dating from 1978 (the latest was published in 1987).

The investigators had access to two groups of adoptees. After searching some 1,646 adoptive records, the investigators came up with 84 adoptive parents who were interviewed. Of the adoptive children, six were considered "definite" or "probable" alcoholics, based on criteria similar to that used in the Danish studies. Three of the six had parents whose adoption records included reference to an "alcohol problem." (There was no way to interview the biological parents and thus no way to apply criteria for alcoholism.) Of the remaining 79 adoptees, only one had a parent with a recorded alcohol problem. Statistically, the difference was beyond chance, but nevertheless the number of subjects was very small.

The six alcoholics were called primary alcoholics, meaning no other psychiatric disorder existed before the onset of alcoholism. Secondary alcoholism refers to alcoholism preceded by depression, antisocial behavior, or other psychiatric conditions (see chapter 3 for more about this distinction). The Iowa group found no correlation between parental alcohol problems and secondary alcoholism in the offspring.

The study involved interviewing adoptive parents. Adoptees themselves were interviewed by telephone, which some critics viewed as an inadequate way to obtain a drinking history. Nevertheless, alcohol problems in the biological parents predicted alcohol problems in the adopted-out offspring, and this continued to be the case when the sample was gradually enlarged over the years.

The alcoholic adoptees were often diagnosed as having behavior problems in childhood. They were still called primary alcoholics, since primary does not preclude the existence of childhood disorders. The latter finding led to increased effort by the Iowa group to study possible interactions between alcoholism and antisocial behavior. Also, like the Cloninger–Bohman group, the Iowa investigators took an increasing interest in the role that environment plays in amplifying or diminishing inherited tendencies toward alcoholism.

In a 1985 study, Cadoret and associates, studying 127 men and 87 women adoptees, found the following: (1) a family history of alcoholism was associated with alcoholism in both men and women adoptees; (2) a family history of antisocial behavior problems was associated with antisocial problems in the adoptees; and (3) there was little overlap between the conditions.

In other words, alcoholism in the biological parents was not associated with antisocial behavior in the offspring and, conversely, antisocial problems in the parents did not predict alcoholism in the offspring. Thus, consistently, the Iowa group found a high degree of specificity of inheritance of antisocial and alcoholic conditions. It was also found that drinking problems in the adoptive home magnified genetic effects, with the men more likely to become alcoholic if they had both a biological and adoptive parent with drinking problems. This finding was confirmed in a 1987 paper by the Iowa group which, again, found that alcoholics beget alcoholics and sociopaths beget sociopaths, but alcoholics do not beget sociopaths and sociopaths do not beget alcoholics.

The Iowa findings were consistent with the original Bohman report of 1978 and also the Danish study showing a high degree of specificity in the transmission of alcoholism. In the Danish study, children of alcoholics had an increased risk of becoming alcoholic when raised by adoptive parents, but did not have an increased risk of having other psychiatric or behavioral disorders, such as drug abuse and antisocial behavior.

The 1987 Iowa study sampled 133 males. The information was obtained in essentially the same way as it had been in the original study. Having interviewed the adoptive parents and talked with the adoptees on the telephone, the Iowa group had a much wider range of environmental variables to analyze than did the Swedish group, who were limited to official records bearing on social class, occupation, and residency. In the 1987 paper, Cadoret analyzed the data using mind-boggling mathematical computations and a computer program.

The findings indicated that biological factors were important in the transmission of alcoholism and antisocial personality, but the transmission was specific for both disorders with little overlap. Cadoret also found that alcohol-related problems in the adoptive family predicted increased alcohol abuse in the adoptees, a finding inconsistent with the Cloninger–Bohman reports and the Schuckit half-sibling study reported in the next section.

In 1986, Cadoret wrote an astute paper on adoption studies and their advantages and deficiencies. Regarding environmental factors, he called attention to the "careful placement" with which adoptees are often matched with foster parents. The adoptive families are investigated and judged to be stable.

This should discourage alcoholism from developing in their adoptive offspring, to the extent that a stable, healthy family protects individuals against genetic inclinations toward alcoholism. Apparently it does not. As shown in the Danish studies, adoptive offspring have about the same rate of alcoholism as children raised by alcoholic biological parents.

As the quotation from Francis Galton at the opening of this chapter eloquently points out, "All the sticks succeed in passing down the current, and they travel, in the long run, at nearly the same rate." The role of nurture is obviously important but difficult to assess. An attempt to do so will be made in the next chapter.

Other adoption studies

In the early 1940s, Anne Roe and her associates at Yale obtained information about forty-nine foster children (both sexes) in the twenty-to-forty-year age group, twenty-two of normal parentage and twenty-seven with a biological parent described as a "heavy drinker." Among the children of heavy-drinking parents, 70 percent used alcohol, compared with 64 percent in the control-parentage group. In adolescence, two children of alcoholic parentage got into trouble because of drinking too much, compared with one in the normal parentage group.

The authors found that adopted children of heavy drinkers had more adjustment problems in adolescence and adulthood than did adopted children of nonalcoholics, but the differences were small and neither group had adult drinking problems. They concluded that there was no evidence of hereditary influences on drinking.

This conclusion, however, can be questioned on several grounds. First, the sample was small. There were only twenty-one male children of heavy drinkers and eleven of normal parents. Second, since women at the time of the study were at very low risk for alcoholism, the discovery that they had no problem with alcohol was not unexpected. Also, the probands in the Roe study differed from the controls in other possibly important regards. For instance, they were older at the time of placement, and a much higher percentage was placed in rural areas or small towns where alcoholism rates were considerably lower than in urban areas.

Another factor, however, may be more germane in explaining the difference between the results of the Danish and Roe studies. Although the biological parents of the proband group in the Roe study were described as "heavy drinkers," it is not clear that they were alcoholic. Most had a history of antisocial behavior, but none apparently had been treated for a drinking problem. By contrast, all of the biological parents of the proband group in the Danish study received a hospital diagnosis of alcoholism at a time when this diagnosis was rarely employed in Denmark, which suggests that only severe alcoholics received the diagnosis.

That the severity of the drinking is an important dimension is further suggested by the finding that only alcoholism, not heavy drinking, differentiated the adopted sons of alcoholics from the adopted sons of nonalcoholics in the Danish study. In short, severe or "classic" forms of alcoholism may have mainly a genetic basis, whereas heavy drinking may have mainly psychosocial origins. (This would be consistent with the Type 1–Type 2 dichotomy suggested by the Bohman–Cloninger study, as well as Cadoret's report that only primary alcoholism ran in families.)

A Danish psychiatrist, Christian Åmark, came to a similar conclusion. In a well-known study of alcoholism in the early 1950s, he reported that "periodic" and "compulsive" alcoholics more frequently had alcoholic children than did alcoholics whose illness presumably was less severe. Home environments were found to be equally good or bad in both groups. Recent studies comparing severe alcoholics with milder alcoholics have found that severe alcoholics more often have a family history of alcoholism. Through the years there have been many attempts to subclassify alcoholism into various types. Based on the preceding observations, it would appear that a classification based on severity and family history might be a useful one.

In 1972, Marc Schuckit and colleagues at Washington University in St. Louis studied a group of individuals reared apart from one of their biological parents where either the biological parent from whom they were separated or a stepparent who raised them had a drinking problem. Strictly speaking, this was not an adoption study, since most of the subjects were raised by at least one biological parent. At any rate, the subjects were more likely to have a drinking problem if their biological parent was alcoholic than if they had a stepparent who was alcoholic. In this study of 32 alcoholics and 132 nonalcoholics, most of whom came from broken homes, it was found that 62 percent of the alcoholics had an alcoholic biological parent, compared with 20 percent of the nonalcoholics. *Simply living with an alcoholic parent appeared to have no relationship to the development of alcoholism.* Cadoret's Iowa studies and the Cloninger–Bohman conception of milieu-limited alcoholism both seem to indicate otherwise—suggesting that postnatal factors do facilitate alcoholism—but the importance of environment remains a subject of debate.

TWIN STUDIES

In addition to studying adoptees, another method for evaluating whether genetic factors may predispose individuals to alcoholism is to compare identical with fraternal twins when at least one member of each pair is an alcoholic. This approach assumes that identical (single-egg) twins differ from fraternal twins only with respect to genetic makeup—that environment is as similar for members of an identical-twin pair as for a fraternal pair. Given these assumptions, the prediction is that genetic disorders more often will occur identically in identical twins than in fraternal twins. Identical occurrence is called "concordance."

The twin approach has been applied to the problem of alcoholism in two large-scale studies—one Swedish, the other Finnish. The Swedish study, conducted by Lennart Kaij, a psychiatrist at the University of Lund, involved 174 male twin pairs where at least one partner was registered at a Temperance Board because of a conviction for drunkenness or some other indication of alcohol abuse. The concordance rate for alcohol abuse in the identical-twin group was 54 percent; in the fraternal-twin group it was 28 percent, a statistically significant difference. Moreover, by dividing alcohol abusers into subgroups based on degree of severity, the largest contrast appeared when the heaviest users of alcohol were considered.

The study also found that social and intellectual "alcoholic deterioration" was more correlated with zygosity (whether the twins were identical or fraternal) than with extent of drinking. (Vagrancy and dementia are major characteristics of alcoholic deterioration.) For instance, a "deteriorated" heavy-drinking identical twin was more likely to have a light-drinking partner who showed signs of deterioration than was a deteriorated fraternal twin. This was interpreted as indicating that alcoholic deterioration occurred more or less independently of alcohol consumption and may be a genetically determined contributor to the illness rather than a consequence.

The Finnish study, conducted by J. Partanen and colleagues, found somewhat more equivocal evidence suggesting a genetic predisposition to alcoholism. The subjects were 902 male twins, both identical and fraternal, 28 to 37 years of age. The sample represented a substantial proportion of all male twins born in Finland between 1920 and 1929. The authors also studied a sample of brothers of the same age group as the twins. Little difference in within-pair variation was found between fraternal twins and nontwin brothers. Subjects were interviewed, and a personality inventory and intelligence tests were administered.

In contrast to the Swedish study, no difference was found between identical and fraternal twins with regard to the consequences of drinking (perhaps the most widely accepted criterion today for diagnosing alcoholism). More or less normal patterns of drinking, however, did appear to reflect genetic factors. Frequency and amount of drinking were significantly more concordant among identical twins than among fraternal twins. Abstinence also was more concordant among identical twins. However, the investigators found no signs of heritability for social complications due to drinking.

The above findings applied to the total Finnish sample. Among younger persons, a difference did exist between identical and fraternal twins with regard to alcohol problems. Identical twins were significantly more concordant for alcohol problems than were fraternal twins.

Throughout this chapter there have been indications from numerous sources that age, severity, and family history are linked. There appears to be a subgroup of alcoholics who become alcoholic at an early age, who have a severe form of alcoholism, and who have a strong family history of alcoholism. The author's favorite term for this group is *familial alcoholism,* but it also has been labeled Type 2 alcoholism, primary alcoholism, and other names. Of all the twin studies—and there are about a dozen—family history is mentioned in only one, and then it is not clear what implications the history has. Twins with a family history of alcoholism might show a very high concordance for alcoholism among identical twins. In twins without such a history, the concordance rate between identical and fraternal twins may be similarly low. Since there is no information about family history, this remains an unsubstantiated theory. The twin studies presumably contain a mixture of familial and nonfamilial twins, and this has conceivably narrowed the gap between identical and fraternal twins in concordance rates for alcoholism.

Meanwhile, there are other twin studies to review. Unlike the Swedish and Finnish studies, they usually were based on records or questionnaires rather than direct interviewing. They include the following:

In 1981, Z. Hrubec and G. S. Omenn studied 16,000 male twin pairs born between 1917 and 1927 who served in the U.S. Armed Forces. The study involved reviewing military service records and Veterans Administration records, and mailing a questionnaire, which elicited a 66 percent return. The findings were as follows: 26 percent of the identical twins were concordant for alcoholism versus 12 percent of the fraternal twins; 21 percent of the identical twins were concordant for alcoholic psychosis versus 6 percent of the fraternal twins; 15 percent of the identical twins were concordant for liver cirrhosis versus 5 percent of the fraternal twins. The differences were all statistically significant. Liver cirrhosis, which is strongly associated with alcoholism, had not been shown heretofore to have a possible genetic predisposition. Although direct interviewing was not done, the study, because of its magnitude, produced valuable findings.

An ongoing study at the famous Maudsley Hospital in London obtained interviews from 56 twin pairs. Questions were designed to determine whether the subjects were psychologically and physically dependent on alcohol. Usual criteria for alcoholism, which include social problems from drinking, were rejected on the grounds that social criteria "reflect public attitudes to excessive drinking rather than the drinking behavior itself."

The Maudsley group found no difference between identical and fraternal twins with regard to concordance for alcoholism, as they defined it. Twenty-one percent of identical twins and 25 percent of fraternal twins were concordant for alcoholism.

Since the Maudsley data differ from most findings in twin studies, the investigators discussed possible reasons for the discrepancy. First, Kaij's Swedish study—the largest study of alcoholism per se—interviewed only men, whereas the Maudsley sample included women. Second, according to the authors, patterns of alcohol consumption in Sweden in the post–World War II period differed from those in England a generation later (how this would contribute to the difference is unclear). Third, and perhaps most important, different diagnostic criteria were used.

Nearly 40 percent of the subjects in the Maudsley sample were below the age of forty. Conceivably, as older twins are studied, findings more consistent with the Swedish study may emerge. The authors believe this is unlikely, and the study stands as a negative report regarding the importance of heredity in alcoholism.

Other twin studies deal with normal and heavy drinking rather than what is clinically known as alcoholism. They include the following:

1. Questionnaires were sent to 850 pairs of same-sex twins chosen from high school juniors who took the National Merit Scholarship test. Included in the questionnaire were thirteen items related to attitudes about alcohol and drinking practices. The investigator, J. C. Loehlin, found that identical twins were more concordant for heavy drinking than were fraternal twins. He conceded that his data were "somewhat fragile."
2. The Maudsley group included five items on alcohol's effects in a questionnaire given to 403 same-sex twin pairs. With regard to more or less normal drinking patterns and psychological effects of alcohol, identical twins were more concordant than fraternal twins.
3. A Swedish group led by R. Cederlof sent questionnaires to some 13,000 twin pairs that included several items about drinking. While there were no differences between identical and fraternal twins with regard to normal drinking, in women identical twins had higher concordance rates for excessive drinking than fraternal twins.
4. E. Jonsson and T. Nilsson studied 750 male twin pairs and found that alcoholic abstinence was under some degree of genetic influence, but identical and fraternal twins had similar concordance rates for how much and how often they drank (which differed from the results obtained in the large Finnish study described earlier).
5. An Italian study of 34 identical and 43 fraternal male twins found no difference between the two groups with regard to amounts of wine consumed.
6. In a Ph.D. thesis published in 1980, N. L. Pederson examined alcohol use in 137 pairs of Swedish twins. He found no evidence for a genetic contribution to beer or wine consumption, but drinkers of spirits seemed influenced by heredity, with identical twins more likely to be concordant than fraternal twins. Pederson, alone among the twin investigators, also addressed the familial nature of drinking practices. The issue

of whether familiar alcoholism was a relevant factor in determining concordance was not clarified.

7. A second Finnish study, conducted by J. Kapprio and associates (the first was conducted by Partanen and associates), examined beer, wine, and spirit consumption of several thousand pairs of twins. As with the first Finnish study, concordance for alcohol consumption varied with age. Concordance for consumption was high in identical twins and low in fraternal twins, but with increasing age the gap narrowed greatly.

In several twin studies, genetic factors appeared to be less important among females than males, but with regard to alcoholism there were usually too few females in the studies to draw firm conclusions.

In summary, a dozen twin studies of alcoholism have been conducted in several countries. Most indicate a genetic influence in both normal drinking patterns and abnormal drinking, with one exception being the preliminary study from London.

THE SEARCH FOR A GENETIC MARKER

An association between alcoholism and other characteristics known to be inherited would afford support for the hypothesis that there is a genetic basis for alcoholism. At least a score of such studies have been done, but the results are inconclusive.

A dozen studies have looked into a possible association between blood groups and alcoholism. About half find such an association, about half do not.

Two studies looked into the possibility that alcoholics might be more or less capable of tasting a particular substance for which our tasting ability is genetically determined. One found more nontasters among alcoholics; one found no difference.

In the mid-1960s a group of scientists in Chile reported that alcoholics were colorblind more often than nonalcoholics. This was followed by several studies in the United States which also reported that alcoholics more often were colorblind, but that the colorblindness was reversible; after several weeks of sobriety, the alcoholics no longer were colorblind, suggesting that the visual defect was due to malnutrition or toxic effects from sustained drinking. Whereupon the Chileans studied the family members of colorblind alcoholics and found a pattern of alcoholism and colorblindness consistent with the inherited form of colorblindness. This study needs to be repeated; if true, it would support a genetic theory. However, there is one strong piece of evidence against colorblindness and alcoholism being co-related on genetic grounds. The gene for colorblindness is located on the X or sex-linked chromosome. The carrier is the mother, who does not get the condition but passes it on to half of her sons. The transmission of alcoholism in families clearly does not follow this progression, but rather appears to be from father to son in most cases.

One other marker study needs repeating. It involves blood groups called ABH secreted in saliva. Some people secrete these blood groups and others do not; genes make the difference.

In 1967, investigators found a remarkable increase in nonsecretors of salivary ABH among alcoholic patients, especially those of blood group A. In 1973, other investigators, studying 448 alcoholics, found the same phenomenon: a strong correlation of alcoholism with nonsecretors of salivary ABH. Again, the association was predominantly in subjects in blood group A.

To be sure, this finding, if confirmed in other studies, may represent an acquired change rather than constitute a genetic marker. Excessive drinking may affect the oral mucosa in some way that blocks the secretion of ABH. Most consider this unlikely. Moreover, why would excessive drinking block secretion in only blood group A alcoholics? The high-risk studies described in chapter 4, in which sons of alcoholics at high risk for alcoholism are studied before they begin drinking, would be an ideal method for exploring the possibility that nonsecretion of ABH is indeed a true genetic marker. As far as I know, this has not been done.

In general, marker studies in alcoholism illustrate what is commonly known as Keller's law (named after the great alcohologist Mark Keller): *"The investigation of any trait in alcoholics will show that they had either more or less of it."* The law applies to much of alcoholism research, but with regard to markers it seems absolutely true.

ANIMAL MODELS

Progress in medical research is often dependent on animal models. Animals have many of the same diseases that afflict humans and can be studied in ways that humans cannot. For example, scientists cannot manipulate human breeding; they can only observe its results. Only animals are suitable for the kind of controlled breeding studies that are required to answer certain questions about genetic transmission.

Until recently there was no animal model for human alcoholism. Some strains of mice and rats drank more than other strains, but never drank like human alcoholics do. To resemble a human alcoholic, an animal would have to do the following:

1. Spontaneously drink enough alcohol to become intoxicated while foods and fluids of equal caloric value were also available.
2. Drink enough to have withdrawal symptoms such as shakes and seizures when alcohol is withdrawn.
3. Drink to relieve these withdrawal symptoms when alcohol is again available.

T.-K. Li and his colleagues at the Indiana University School of Medicine have developed animals that do all of the above. They took rats that drank more than other rats (although not to the point of intoxication) and inbred them over a number of generations. They also took rats who abjured alcohol and inbred them. When enough generations had passed, they had a group of heavily drinking rats and another group of teetotalers.

The drinking rats were alcoholic by almost any definition. They voluntarily drank large quantities of alcohol. They drank enough to become drunk. They worked to obtain alcohol by pressing a lever, even when food and water were available. They became physically dependent on alcohol and had withdrawal symptoms. They became tolerant of alcohol so that they could drink more and more while showing the effects less and less.

Finally, they seemed to "enjoy" alcohol. Originally there was some question about whether the rats drank because they liked the taste or smell of alcohol. However, they preferred alcohol to water even when it was injected directly into their stomachs. It took some imagination to show this. The investigators used an apparatus that automatically delivered either water or an alcohol solution directly into the stomachs of alcoholic and teetotaling rats whenever they drank either of two water solutions flavored with almond or banana. The idea was to train the animals to associate a particular flavor in their drinking water with the presence or absence of a pharmacologic effect from the stomach alcohol infusion, without giving them the opportunity to taste or smell what was entering their stomachs.

It worked. The alcoholic rats drank fourteen times more of the flavored water that was linked to alcohol infusion than did the teetotaler rats. In fact, the rate of consumption was higher than when they drank the alcohol, suggesting the taste of alcohol may have been unpleasant. They achieved blood levels of alcohol four times higher than the legal driving limit in most states!

A tendency to relapse is a major feature of alcoholism in humans. Alcoholic rats also show the tendency. After a few days on the wagon (so to speak), they rapidly return to heavy drinking when alcohol is made available.

It may be going a little far to say that some rats enjoy drinking. However, they appear to drink alcohol for the same reasons that many people drink: for its pharmacologic effect on the central nervous system.

If rodents can inherit alcoholism there is little reason to

doubt that humans can too. But heredity obviously is not the whole story. Alcoholism has many aspects that cannot be explained by genetics. These will be discussed in chapter 6.

THE PROTECTION FACTOR

Speculation about genetic factors in alcoholism is not complete without commenting on physiologic reactions that deter, or protect, large numbers of people from becoming alcoholic.

For example, many women are intolerant of alcohol. After drinking small amounts they experience unpleasant physiologic reactions, ranging from mild nausea to dizziness or somnolence. This may explain partly the fact that fewer women than men are alcoholic. However, the role of physiologic protective factors in deterring heavy drinking is best documented by studies of the so-called Oriental flushing phenomenon.

As described earlier, roughly two thirds to three quarters of Orientals experience a cutaneous flush, mainly of the face and upper part of the body, after drinking a small amount of alcohol (e.g., fewer than one or two beers). The flush is usually accompanied by unpleasant subjective reactions, including feelings of warmth and queasiness, as well as an increase in heart rate and decrease in blood pressure. The Oriental flushing phenomenon undoubtedly is genetic in origin. Oriental infants have been given tiny amounts of alcohol and they too have flushed. Some studies, not all, find elevated levels of acetaldehyde correlated with the flush, and atypical forms of both alcohol dehydrogenase and aldehyde dehydrogenase (enzymes that metabolize alcohol) have been reported in a high proportion of Japanese men and may presumably be related to the flush.

Whatever the biochemical mechanism, the flushing reaction undoubtedly constitutes a powerful deterrent to heavy drinking for many Orientals. The reported low alcoholism rate in much of the Orient thus may have a genetic explanation, reinforced perhaps by cultural factors.

CHAPTER SIX

The Case for Nurture

> To drink is a Christian diversion
> Unknown to the Greek and the Persian
> —William Congreve

Any final explanation for alcoholism must explain more than why alcoholism runs in families. It must explain the following:

Why is there a low rate of alcoholism among Jews and a high rate among the Irish?

Why are there more French alcoholics than Italian? Why are there more alcoholics in northern France than in southern France?

Why is alcoholism rare in the Orient?

Why is alcoholism more common in cities than in rural areas?

Why are more reporters alcoholics than mail carriers? More bartenders than bishops?

Why are more men alcoholics than women?

Imagine a hereditist and an environmentalist (perhaps over a drink) discussing these statements point by point. Their argument might be as follows:

H: Jews intermarry. Hispanic Jews, for example, are unusually susceptible to a blood disease known to be inherited. Why is it not reasonable to suppose that Jews would be "protected" from developing alcoholism by a similar hereditary factor?

E: Hispanic Jews do intermarry, but this is much less true of Jews in other places. In America, for example, Jews frequently marry non-Jews, and it is inaccurate to speak of a genetically pure Jewish "race." Nevertheless, despite intermarriage with other ethnic and racial groups and increasing assimilation of Jews into the general American culture, their alcoholism rate remains low. If heredity cannot account for this, it must be attributed to environmental factors. Even with assimilation, Jewish tradition continues to influence the way Jewish mothers and fathers rear their children. Traditionally, moderate drinking is condoned, but drunkenness, particularly chronic drunkenness, is condemned and viewed not only as an individual weakness but a family disgrace. In families where some Jewish tradition persists, closer family ties prevail than generally exist among non-Jews. These factors provide a more plausible explanation for the low alcoholism rate among Jews than does any genetic explanation.

E *(continuing)*: As for the Irish, no one ever accused them of being a race. A very high proportion of Irish have emigrated to other countries where intermarriage with people of other ethnic and racial backgrounds has been widespread. Again, family and cultural traditions provide a more plausible explanation for high rates of alcoholism among the Irish than heredity does.

E *(continuing)*: As for the French and the Italians, it is true that people from the two countries differ to some extent in physical characteristics, but there is no known instance of people living in France being more prone to a particular genetic disorder than people living in Italy. Again, there is frequent intermarriage with people from other countries. If indeed there are more alcoholics in northern France than in southern France, this would further support the importance of environmental factors. Perhaps climate is a factor. Since the north is chillier than the south, perhaps northerners drink more to keep warm. It sounds farfetched but no more so than a genetic explanation, considering that northerners and southerners regularly intermarry.

H: Three points need to be made. In the first place, are the French really more often alcoholic than Italians and does alcoholism really occur more often in northern France than in southern France? True, this has been reported, but estimates of the prevalence of alcoholism are notoriously unreliable. They are usually derived from cirrhosis figures or admissions to hospitals for alcohol problems. Cirrhosis figures themselves are not very reliable, and whether a person becomes "officially" an alcoholic by receiving treatment for the condition is influenced by many factors other than the presence of alcoholism itself. One is whether facilities are available for treatment. Another is the readiness of individuals to seek treatment. Another is the diligence with which public officials maintain treatment records. Perhaps in northern France the authorities keep better records than in southern France. This could entirely explain the alleged difference.

At any rate, some people have challenged the repeated assertion that alcoholism is more prevalent in France than in Italy, as indeed some have questioned whether alcoholism is all that common in Ireland. Everyone agrees that Irish immigrants in the United States seem particularly susceptible to alcoholism. Ireland itself, however, has a fairly low reported rate of alcohol-related liver disease, and some people question whether alcoholism is any more common in Ireland than, say, in England. Perhaps in rural areas it may be less common if only because rural Ireland is less affluent and people have less money to buy alcohol.

This brings us to the second point: Alcoholism can only occur when alcohol is available. It is often said that prohibition in the United States was a failure. In some ways, it was a huge success. Alcohol-related liver-disease rates dropped precipitately, as did admissions to hospitals for alcohol problems. Similarly, in France during World War II, when wine was rationed, alcohol-related liver disease became less common. In Scandinavia there has apparently been some success in reducing alcohol consumption and presumably alcoholism by increasing the cost of distilled spirits. Perhaps, if it is true that Italy has less alcoholism than France, it is partly because Italy is a poorer country and fewer people can afford distilled spirits.

Finally, even if these differences between cultures and countries are genuine, what we call alcoholism may be a mixed bag of conditions. Some kinds of alcoholism may be genetically influenced while others are not. Because of varying definitions of alcoholism and unreliable means of estimating the prevalence of alcoholism, it is difficult to say what proportion of alcoholics have what might be called "familial alcoholism" as opposed to alcoholics whose excessive drinking has other causes.

H *(continuing)*: As for the rarity of alcoholism in Oriental countries, again one might question the accuracy of the observation and also raise the possibility that officials in such countries as Communist China might try to conceal whatever alcoholism exists so as to make their system of government look good (as, indeed, occurred until recently in Russia). Assuming, though, that alcoholism is rare in most Oriental countries—and the assumption seems more justified than most—here we have an instance where there is a known physiological difference between one race and other races in their response to alcohol. Several studies have indicated that many Orientals have a low tolerance for alcohol. After drinking a small amount, they flush, develop hives, feel nauseated, and generally are discouraged by their physical reaction to alcohol from drinking more. It is unlikely that these responses are produced by psychological factors arising from social pressure; there are fewer taboos against drinking in the Orient than there are in the American Bible Belt. The low rate of alcoholism in the Orient, in other words, may really reflect a biologically determined low tolerance for alcohol.

E: There are also many *non*-Orientals who have a low tolerance for alcohol. Relatively few may develop skin flushing, but many people, after a few drinks, become dizzy, develop a headache, or feel sick. Thus large numbers of non-Orientals also are deterred by their physical response to alcohol from drinking excessively. Nevertheless, this leaves considerable numbers of people without low tolerance and from this group America alone produces several million alcoholics. The proportion of Orientals who cannot drink much because of adverse effects from alcohol may be higher, but many can drink as much as they please, and since alcohol is ubiquitous (all that is required is yeast, sugar, and water), one would expect higher rates of alcohol problems in the Orient than apparently exist, unless social factors discourage alcoholism.

E *(continuing)*: There is another point about the importance of culture and tradition. Anthropologists have observed that every society tends to have its drug of choice, its "domesticated intoxicant" that is favored by the great majority of people over other intoxicating substances. In China, for example, the favored intoxicant for centuries was opium. In India, the Middle East, and North Africa cannabis derivatives (hashish, marijuana) historically have been the most widely used intoxicant. In Judeo-Christian societies alcohol has been the drug of choice.

In recent years, marijuana has become a competitor to alcohol among the younger members of Western countries. This cultural heresy, as it were, originally was met with strong legal and moral resistance, which has abated considerably in the past ten years; nevertheless, whether marijuana in the West is a passing fad or will remain popular is uncertain. In the past, attempts by other societies to have more than one "approved" intoxicant have ultimately failed. An example is the religious taboo against alcohol among Moslems, a taboo coupled with religious and legal sanctions as stern as the long prison sentences until recently meted out to marijuana users in this country.

Now it is very difficult to explain these cultural differences in intoxicant choice to inherited factors. It is more logical to attribute them to historical accident that led to the selection of one substance and to the development of customs that control its use with minimal damage to society. The "accident" may have been nothing more than the availability of the substances, such as cannabis in India, the poppy in China, and cocaine from the coca leaf in the Peruvian Andes.

H: All this is true, but it does not explain why certain users of these substances in every society get into serious trouble from their use while apparently the majority use the substance with little or no ill effects. This raises an important scientific question: Are some individuals born (if you'll pardon the expression) with a nonspecific vulnerability to addictiveness—with the substance to which they become addicted determined by availability and cultural sanction—or is addiction-proneness specific for certain substances? In other words, is it possible that some individuals have a specific vulnerability to alcohol abuse but can use opium products, nicotine, and other addictive substances within moderate bounds?

The question is still open. Favoring a nonspecific addiction proneness is the fact that alcoholics almost universally are heavy smokers and that heroin addicts, if denied heroin, often become alcohol abusers. It would be interesting to know whether opium and cannabis abuse runs in families, as alcohol abuse does. At the moment there is little information on the subject.

Nevertheless, while it is true that there are cultural variations in the preference for intoxicating agents, it is also true that usually only a minority of individuals exposed to the agent develop serious problems from its use, and that just as the hereditist must explain cultural differences in alcohol use, environmentalists must explain why only a minority of users become abusers.

H *(continuing)*: This brings us to the next question: Why is alcoholism more common in cities than in rural areas, more common in some occupations than in others, and more common in men than in women?

No one questions the importance of custom and tradition in determining how many people use a particular intoxicating substance such as alcohol. Until recently in the United States, for example, there were social sanctions against women drinking at all, much less becoming intoxicated. There have been religious sanctions against any drinking at all in rural areas where fundamentalism is strong. If a person does not drink, he or she cannot become alcoholic. As society relaxes its pressure concerning the use of alcohol by women (with women freely permitted to go to bars alone and so forth), it is quite likely that the number of alcohol abusers among women will increase (and there is evidence that this is happening).

There is also some evidence that the proportion of abusers to users remains constant. In other words, if the abuse potential for alcohol is, for example, one in ten, 10 percent of men and women drinkers will be alcoholic. As the number of women drinkers increases, so also will the number of women alcoholics. This, again, indicates that factors other than social and cultural ones—which no doubt contribute to whether a person drinks or not—are operative in selecting those who ultimately will drink too much. The "selector" may well be a biological, genetically determined susceptibility.

H *(continuing)*: As for occupational differences in alcoholism rates, they can be explained fairly easily. Some occupations are more tolerant than others of drinking and heavy drinking. For example, by the nature of his work, a reporter or bartender or housepainter has more freedom to drink than does a chest surgeon or an airline pilot (both of whom risk censure or worse just by having alcohol on their breath). People who are inclined to drink heavily very likely gravitate to occupations where heavy drinking is tolerated—if not actively encouraged (as is the case in public relations with its tradition of the two- or three-martini lunch). People rarely go into various jobs purely by chance. Of the many factors that influence job choice, one, for some, may be the opportunity it provides to drink. This is an instance where tradition and opportunity open the door for genetic traits to walk in. It also shows how group differences in alcohol use, ostensibly resulting from nurture, may actually be nature having her way incognito.

E: William James said something like the following: Toss down a handful of beads and, by ignoring certain beads, you can perceive any pattern you wish. In looking for a genetic explanation for cultural variation in drinking patterns, it seems that is precisely what you are doing. In your review of the "case for nature" (I assume the last chapter was yours), a progenetic bias also seems obvious. Let me quickly summarize the studies presented in the last chapter.

The recent adoption studies are fairly consistent. Sons of alcoholics often become alcoholic even when raised by adoptive parents. It is less clear whether this occurs also in daughters. And there is an earlier adoption study, by Anne Roe, that doesn't show an association between parental alcoholism and drinking by adoptees in either sex. Devoted hereditists either ignore this study or explain away its findings. (The hereditist who wrote chapter 5 relegated the study to a section called "other studies.") The Roe study is flawed, but so are all the adoption studies. One clearly sees a double standard lurking here.

The twin studies are a mess. Only two of them really address alcoholism—Kaij's in Sweden and the Maudsley study. The Swedish study found evidence favorable for a genetic hypothesis. The Maudsley study found the reverse. Which should one believe? Granted, the two studies define alcoholism somewhat differently, but this would hardly seem to explain the discrepancy. Most of the other twin studies relied on questionnaires which included a few items about drinking buried in a mass of questions about other matters. One study used telephone interviews! Do you really think anyone with a drinking problem would confess this to a stranger by telephone? It's hard enough to get people to admit to a drinking problem face to face or tell their own trusted doctor. Questionnaires and telephone calls are absurdly inadequate methods for obtaining such sensitive information. Nevertheless, when the information obtained fits the hereditist's bias, it is usually accepted as incontrovertible truth.

The genetic marker studies—attempts to associate alcoholism with a known inherited trait—are even more chaotic than the twin studies. For every study that reports an association with a particular marker there is at least one other that finds no association or has a nongenetic explanation.

As for the animal studies, Dr. Li's success in creating "alcoholic" rats was interesting. However, he produced them in a way that fits no conceivable human circumstance: inbreeding for a single variable over many generations. There is something called assortive mating in humans—a tendency for people to marry people of similar interests and background—but this is not even remotely comparable to multiple generation inbreeding of a type most commonly identified with horticulture.

Moreover, drinking behavior by rodents is not entirely controlled by genes. This is demonstrated by at least two studies. Both involved two strains of mice. One strain spontaneously drank more alcohol than the other. Everyone assumed this was because of genes. Then the investigators did an "adoption" study. They took the babies of the high-preference strain and "adopted" them out to female mice of the low-preference strain. They also did the reverse, having newborn mice of the low-preference strain raised by females of the high-preference strain.

The results were interesting. The baby mice from the high-preference strain reared by mothers of the low-preference strain spurned alcohol; baby mice of the low-preference strain reared by mothers of the high-preference strain had an increased preference for alcohol.

How can this be explained? The answer is that even baby rats learn from their mothers.

E *(concluding)*: Let me summarize what I believe are the strongest arguments on both sides of the issue. The strongest evidence for heredity is that alcoholism runs in families, even when the children are separated from the alcoholic parents and raised by adoptive parents. The strongest evidence that social factors contribute to alcoholism is the great diversity in alcohol use and alcohol abuse among various cultures, nations, ethnic groups, social classes, regions, sexes, and other groupings. With enough ingenuity, the hereditists can see heredity rearing its head even in these domains, but for the scholar without a progenetic bias, attributing these differences to heredity seems farfetched.

THE CRUCIAL QUESTION

As our two polemicists finish their argument and their drinks, pay the bill, and depart into the night, let us repeat a question already raised by the hereditist and deal with it as it might be dealt with by the environmentalist. It is the crucial question. In this country most adults drink (roughly 70 percent) and one out of every twelve or fifteen becomes alcoholic. Why are children of alcoholics more vulnerable to alcoholism than children of nonalcoholics?

Here is some nongenetic conjecturing:

1. *Alcoholics make bad parents.* H. J. Clinebell, a minister who has counseled many alcoholics and their families, lists four ways in which they make bad parents. First, when one parent is alcoholic, there may be a shift or reversal in the parents' roles, complicating the task of achieving a strong sense of sexual identity in the children. Second, an inconsistent, unpredictable relationship with the alcoholic parent is emotionally depriving. Third, the nonalcoholic parent is disturbed and therefore inadequate in the parental role. And fourth, the family's increased social isolation interferes with peer relationships and with emotional support from the family.

The big problem with the bad-parent explanation is that there are far more bad parents than there are alcoholics. Many children are raised by only one parent or by no parents. With our current high divorce rate, broken homes have become increasingly common. It is unlikely that alcoholism rates have increased proportionately.

Proponents of the bad-parent theory usually deal with these arguments by saying that alcoholism springs from an upbringing that was bad in a *specific* way. However, there is no agreement about what the specific way is.

2. *Children learn from their parents.* This explanation works in two ways. Observing the havoc produced by their alcoholic parent (or parents), children resolve never to touch alcohol and become adamant teetotalers. Undoubtedly this occurs, but apparently not very often.

The reverse is that children model themselves after their parents; their father or mother (or both) drinks and therefore they drink.

"Dad was only really happy when he drank," observed the son of an alcoholic. "Sure, he was sometimes mean when he drank and often an embarrassment to the family. Nevertheless, when he was sober he always seemed tense, withdrawn, unhappy. I much preferred to see him drink, if it wasn't too much. He became jovial, interested in the children, accessible. I came to identify good feelings and a warm and sociable manner with drinking."

Parents also transmit tradition. If they speak English, their children speak English. Social attitudes toward drinking pass from parents to children, who acquire them as naturally as they do their parents' accents and table manners.

Social values, including those regarding alcohol, find expression in some families more so than in others. A good example is the family of Eugene O'Neill, the playwright. When O'Neill began drinking, according to biographers, he simply was following the example of his father and brother. His father drank daily, usually in barrooms. "You brought him up to be a boozer," says Eugene's mother to his father in *Long Day's Journey into Night.* "Since he first opened his eyes, he saw you drinking. Always a bottle on the bureau in the cheap hotel rooms!"

To some extent, O'Neill's father was a conduit for Irish attitudes toward alcohol prevailing at the turn of the century. When Eugene was an infant and had a nightmare or stomachache, his father gave him a few drops of whiskey in water. O'Neill later believed this old Irish custom contributed to his drinking problem as an adult.

"I'm all Irish," O'Neill said, referring not only to his ancestry but also to the Irish customs and attitudes of his family, exquisitely portrayed in *Long Day's Journey into Night.* In the play, as John Henry Raleigh pointed out, a bottle of whiskey is at the center of the room, in many ways its most important object. "If not using it they talk about it. It enters into their very character; the father's penuriousness is most neatly summed up by the fact that he keeps his liquor under lock and key and has an eagle eye for the exact level of the whiskey in the bottle . . . by the same token, the measure of the sons' rebellion is how much liquor they can 'sneak.' "

Raleigh wrote that "sneaking a drink" had more significance for the Irish than for other cultural groups. "Allied to this peculiar Irish custom are the concomitant phrase and action: watering the whiskey, that is filling the bottle with water to the level where it was before you sneaked your drink." In some Irish households, Raleigh said, "whole cases of whiskey slowly evolve into watery, brown liquid, without the bottles ever being set forth socially, so to speak. This act—the lonely, surreptitious, rapid gulp of whiskey—is the national rite . . . "

National or not, it was clearly an O'Neill family rite, as was the medicinal use of alcohol, another custom traditionally Irish.

"The Irish addiction to drink is a simplifying element in their lives," wrote Raleigh. "This is how all problems are met—to reach for the bottle." And reach the O'Neills did. "A drop now and then is no harm when you are in low spirits or have a bad cold," advises the maid in *Long Day's Journey,* and Eugene's father agreed. "I've always found good whiskey the best of tonics," he says in the play, calling drink the "good man's failing." Even O'Neill's mother found alcohol a "healthy stimulant."

According to Irish scholars, alcohol served utilitarian functions other than medicinal, at least in earlier times. The Irish were as Puritanical about sex as they were tolerant toward drunkenness, and in fact the two were linked—alcohol served as a sexual substitute. The teetotaler, indeed, was considered a menace, a man who prowled the streets and got girls into trouble. When a young man was unhappy in love, he was advised to "drink it off." Wrote one sociologist: "Drowning one's sorrows becomes the expected means of relief, much as prayer among women."

These attitudes toward alcohol, held by at least some Irish families at the turn of the century, are a far cry from attitudes toward alcohol held by Southern Baptists or Jews. The former condemn any use of alcohol and the latter condemn drunkenness.

How these traditions develop in a given society is as complicated and hard to comprehend as is the pairing and unpairing of molecules in the microscopic world of genes. But obviously customs toward drinking, like "prayer among women," do not easily lend themselves to a genetic explanation.

3. *Frustrated, unhappy, insecure, lonely people drink to feel less frustrated, unhappy, insecure, and lonely.* This, perhaps, is the most popular explanation of why people drink to excess. The assumption is that people drink to escape. They have stresses to which they respond by losing themselves in alcohol.

The fallacy in this reasoning is a simple one: Most people are frustrated, unhappy, insecure, or lonely much of the time but do not become alcoholic. Thirty percent of adults do not drink at all. This is not a purely environmental explanation in any case. The question has to be asked: Why do some individuals respond to stress more intensely than others? Response to stress itself may in part be genetically determined.

The same point can be made about people who drink apparently to relieve depression. Certain depressive illnesses themselves may be genetic; there is substantial evidence for this. Moreover, most people probably do *not* drink to relieve depression. In a study of patients with depressive disorders, one third drank somewhat more when depressed but another third drank somewhat less. In fact, depression may follow rather than lead to drinking. In studies where individuals have been given large amounts of alcohol, they develop depressive symptoms.

4. *Alcoholics are loners.* In a letter to the author, the historian Dr. Gilman Ostrander proposed the following explanation for alcoholism:

> Alcoholism is basically a disease of individualism. It afflicts people who from early childhood develop a strong sense of being psychologically alone and on their own in the

> world. This solitary outlook prevents them from gaining emotional release through associations with other people, but they find they can get this emotional release by drinking. So they become dependent on alcohol in the way other people are dependent on their social relationships with friends and relatives.

Ostrander believes his theory explains why alcoholism is more prevalent in some ethnic groups than in others. The high alcoholism rate among the Irish and French, he says, is at least partly traceable to the fact that Irish and French children are brought up to be

> . . . responsible for their own conduct. When they grow up and leave the household, they are expected to be able to take care of themselves. Individualism in this sense is highly characteristic of these groups.

Jews and Japanese, on the other hand, have a low alcoholism rate because children are *not* expected to be independent.

> Infants in these families are badly spoiled, that is to say, their whims are indulged in by parents and older relatives, so that, from the outset, they become emotionally dependent upon others in the family. . . . It is never possible for them to acquire the sense of separate identity, apart from their family, that is beaten into children in, say, Ireland . . . They are likely to remain emotionally dependent upon and a part of their family in a way that is not true in societies where the coddling of children is socially disapproved of.

And this is why Ostrander believes most Jews and even hard-drinking Japanese do not become alcoholic.

> They never had the chance to think of themselves as individuals in the Western sense of the word. They are brought up to be so dependent upon others in the family that they are unable to think of themselves as isolated individuals.

It is a hard theory to prove. There may even be some question about the low rate of alcoholism in Japan relative to other Far Eastern countries. Nevertheless, it is an interesting explanation for the differences that appear to be real between various cultures, and deserves attention from sociologists who might find ways of testing the theory scientifically.

In conclusion, it should be noted that nobody disputes the importance of cultural factors in drinking behavior, and it seems entirely possible, though difficult to prove, that environmental factors have some bearing on whether a person becomes a problem drinker or an alcoholic. At the very least, alcohol must be available for alcoholism to occur; a person must drink before he can be a drunkard. Whether or not he drinks certainly will be influenced by his social environment. Whether he drinks excessively also may be influenced by his social environment. The puzzle remains: What specific factors—genetic, environmental, or both—produce serious alcohol problems in a rather small minority of drinkers? There are still research leads to be explored, and they will be discussed in the next chapter.

THE COA MOVEMENT

Beginning in the early 1980s a remarkably new conception and concern arose on the medical-social horizon. The concern centered on a sizable group of people called Children of Alcoholics (COAs).

There were an estimated 28 million COAs in the country—one out of every eight Americans. Conferences sprang up, well attended by social workers, alcoholism counselors, COAs, and a smattering of psychologists and psychiatrists. A national COA foundation was established. A new lecture circuit came into existence starring COA experts, who often were COAs themselves.

The COA movement was mainly promoted by people who adopted enviromentalist explanations for alcoholism. Behind the movement is the explicit assumption that children are mainly products of their upbringing. In a sense, the COA movement has been a counterweight to the genetic studies—keeping the nature–nurture controversy in a dynamic equilibrium.

Congress held hearings about COAs, sometimes seeming to elevate the COA problem to the level of the AIDS problem. There was much impassioned testimony. The influential New York lawyer Joseph Califano testified that "of all the major diseases, only heart attacks and strokes come close to alcohol abuse in cost." COAs, he said, are "likely to be victims of abuse, be expelled from or drop out of school and have migraine headaches, asthma, and other health problems." COAs are "prone to depression, introversion, schizophrenia . . . and bulimia." Even COAs who appear to do well have secret problems. Califano, an honorary trustee of the Children of Alcoholics Foundation, was concerned about the children becoming "super-copers" who "took responsibility for the family by caring for others, cooking meals, running the household, and even balancing the family checkbook, who later become super-overachieving executives, prone to serious problems in the workplace."

On the psychiatric scene, COAs became identified as a unique group with specific problems requiring special care. In some hospitals, wards were set aside just for COAs. In 1987 the COA Foundation identified 235 programs for COAs in 34 states. Still, the new COA experts were concerned that most victims were not being reached. About 75 percent of the programs had been in operation for three years or less; they tended to be small, serving fifty or fewer children a year, and they were concentrated primarily in California, New York, Wisconsin, and Massachusetts. "The established service delivery systems are generally not cognizant of or organized to respond to the needs of children of alcoholics," according to a COA spokesperson.

But is it true? Are children of alcoholics particularly prone to specific disorders requiring specialized care? Based on review of the large body of literature on the subject, the answer is unproven. Most of the literature consists of testimonial and anecdotal reports from patients or of such groups as Al-Anon. Reviewing this literature, one finds that children of alcoholics are poor students or drug abusers; are hyperactive and antisocial; make frequent suicide attempts; have high rates of psychiatric illness, and are often alcoholic.

The last part about being often alcoholic is indisputably true. As noted frequently in this book, the risk of alcoholism in the children of alcoholics is four- or fivefold that of the risk in the general population. Moreover, the risk remains about the same whether the children are raised by their alcoholic parents or by nonalcoholic adoptive parents. In other words, heredity as well as environment may explain the familial link.

One common theme emerges from the wealth of anecdotal reports. Alcoholic parents are unpredictable. They sometimes come home and sometimes do not. Their behavior is erratic. The kids hate to bring friends over because they cannot be sure how their parents will behave. The family becomes isolated from neighbors and other families because Mom or Dad would rather drink and serious drinking is basically a solitary business.

So the children indeed become "super-copers" and often assume the parental role in the ways mentioned by Califano. Although Califano is concerned that some of these super-copers end up as super-achieving, overstressed executives, a case could be made that assumption of responsibility early in life builds character and prepares children for the coping ahead of them. There are certainly many instances of successful, admired people who as children lost their parents by death and became heads of families at a young age. In previous centuries, when parents commonly died young of illness or in childbirth, fourteen-year-olds assumed parental responsibility, often without apparent ill effect. Of course, they did not suffer the humiliation of having friends see a mother or father under the influence of drink, but there were many other serious matters to cope with, such as a foreclosed mortgage and foxes in the chicken pen.

There are only a few studies where children of alcoholics are compared to other children. They generally show that children of alcoholics have more psychological and behavioral problems than children of nonalcoholics, but not many more. One study shows that children of alcoholics have more problems but no more than children of a parent who suffered a serious depression.

There is another source of data about children of alcoholics that also raises doubt about the theory that they represent a special group. Chapter 4 reviews high-risk studies in which children of alcoholics are compared to children of nonalcoholics. The former sometimes do report more hyperactivity and behavioral problems in school. However, there are no consistent reports that children of alcoholics differ from children of nonalcoholics with regard to mental and emotional problems. Differences may exist but, if so, they have been overlooked.

High-risk studies involve nonpatients. Biases may enter into the selection process, but patient status is not one of them. High-risk studies should probably be considered more reliable than case reports.

Whether or not COAs represent a special psychiatric group, the COA movement has accomplished one worthwhile goal: More children of alcoholics are learning that their chances of becoming alcoholic are greater than if they were children of nonalcoholics. Whether or not heredity is the reason, alcoholism clearly runs in families, and families should know about this. Oddly, it seems, many families do not. In 1987, a telephone poll in New York State revealed that only 5 percent of respondents knew that alcoholism ran in families, including members of alcoholic families.

The knowledge, one hopes, will soon be universal. There are instances of people who watch their drinking very carefully because they have alcoholism in the family. Just as children of diabetics should have their urine checked occasionally for sugar, children of alcoholics should be aware of their vulnerability. Knowledge is power.

THE DEBATE GOES ON

Even though it seems clear that, in a practical sense, people should probably act as though alcoholism *is* hereditary, a conclusive answer to the question posed by the title of this book remains elusive. The nature–nurture debate continues, with environmentalists and hereditists often agreeing on facts while disagreeing about their meaning.

The history of this debate is long and spirited. For example, as the 1944 session of the Yale Summer School of Alcohol Studies ended, the famous E. M. Jellinek was asked the following question:

"Sir, you referred to a sample of over 4,300 inebriates, of whom 52 percent came from alcoholic parents. What was the extent of the alcoholism of the parents?"

Jellinek replied: "The alcoholism in those parents was real honest-to-goodness inebrity . . . but this does not mean the alcoholism was transmitted biologically. It was transmitted socially."

How did this least dogmatic of alcohologists come to such a dogmatic conclusion? He never said. However, his other comments that day suggest several possibilities.

Jellinek recognized that alcoholism ran in families but was equally impressed by how many families it did not run in. This suggested to him that, "if a hereditary constitutional factor is present, it does not become operative without intercurrent social factors."

Second, Jellinek did indeed dislike dogmas. One nineteenth-century dogma that still lingered was that alcoholism *was* hereditary. Since there was almost no evidence for this except for the "familialness" of the condition, Jellinek rebelled, although his rebellion might have gone a little too far.

Third, Jellinek fell back on a non sequitur that is still heard on the alcoholism lecture circuit: If alcoholism is hereditary, it cannot be treated. If it can be treated, it must not be hereditary. Jellinek surely saw the illogic of this, but still believed that therapists should not be thwarted by "implacable fate," that is, heredity.*

(This is nonsense, of course, since many hereditary illnesses are treatable, and treatability is not relevant to the issue of origin in the first place.)

Still, Jellinek knew the right questions to ask. If anything is inherited, what is it? Does it involve tolerance? Are there internal "requirements" that must be present for alcoholism to appear? Does a predisposition to alcoholism, a "readiness to acquire the disease," reflect a specific proneness to alcohol abuse—or general psychological proclivities that favor the development of alcoholism if no other "escape" is available?

When Jellinek discussed the subject in 1944, only one nature–nurture study existed, and it wasn't published until a year later. Even in 1976, when the original version of this book was published, there was a short supply of evidence supporting a role for heredity in alcoholism. That first version was on the slender side, and the title ended with a question mark. Today, it is fatter, much fatter. Much has happened over the last eleven years, but the question mark remains. The need for that punctuation may disappear in time for the next revision. Since then, more evidence that alcoholism is a family disorder has accumulated. One reviewer found 140 studies reporting on the high prevalence of alcoholism in the families of alcoholics.

Not everything that runs in families, however, is hereditary. For example, speaking French runs in families but is not hereditary.

Before predicting any breakthrough, we should remember the story of *kuru:*

*Jellinek's mixed feelings about alcoholism and heredity were evident when he came up with a second non sequitur of even greater beauty: "The only permissible conclusion is that not a disposition toward alcoholism is inherited but rather a constitution involving such instability as does not offer sufficient resistance to the social risk of inebrity." No better example of having one's cake *and* eating it can be found in the literature about alcoholism.

Kuru is a disease limited to a small native population on a South Pacific island. Like multiple sclerosis, it destroys the nerve sheaths. It runs in families, from fathers to sons, and for a while no one doubted that it was inherited.

Well, someone doubted it (probably a sociologist). The natives were cannibals. When their fathers died, the sons, following tradition, ate their fathers' brains.

The brains, it seems, can contain a slow virus—one that takes years to show its presence by destroying the nerve sheaths. The virus, not genes, was the culprit—and viruses are environmental. Kuru wasn't hereditary after all.

That's why the question mark remains after *Is Alcoholism Hereditary*?

Who knows? Perhaps, someday, speaking French will turn out to be inherited.

CHAPTER SEVEN

What Next?

> "Drinking when we are not thirsty and making love at all seasons, Madam; that is all there is to distinguish us from the other animals."
>
> —Pierre de Beaumarchais

In science the question is as important as the answer. Some questions inherently are unscientific, meaning they cannot be answered in such a way that scientists would agree upon the answer. "Does God exist?" is widely considered such a question. It has been said that science can deal only with mundane questions, such as how many legs a horse has, or other questions that can be solved by counting, whereas the important questions, such as "*Why* does a horse have four legs?" or the one about God are beyond the reach of science. This may be true, but nevertheless, if science is viewed in its simplest terms as a search for reliable information, scientists are probably worth their keep (just as detectives and internal-revenue agents are probably worth theirs).

In deciding what problems are suitable for scientific inquiry, four things need to be known (insofar as possible):

1. What is the question being asked?
2. Is the question answerable?
3. What will it cost to answer the question?
4. Is it worth the price?

Whether a particular answer is worth the price is something for society to judge. The scientist's job is to produce questions that are answerable and let the taxpayers and bureaucrats decide whether the price is right (the exact price, of course, cannot always be known in advance).

Here are some questions about alcohol and alcoholism that research may answer:

1. Why do people drink?
2. Why do some people drink more than others? Why *can* some drink more than others?
3. Why do some drink to the extent of causing serious harm to themselves and others?

QUESTION 1: WHY DRINK?

At one level, the answer to the first question seems obvious. Berton Roueché expressed it as well as anyone:

> The basic needs of the human race, its members have long agreed, are food, clothing, and shelter. To that fundamental trinity most modern authorities would add, as equally compelling, security and love. There are, however, many other needs whose satisfaction, though somewhat less essential, can seldom be comfortably denied. One of these, and perhaps the most insistent, is an occasional release from the intolerable clutch of reality. All men throughout recorded history have known this tyranny of memory and mind, and all have sought, and invariably found, some reliable means of briefly loosening its grip.

The answer, while eloquent, is unscientific. No answer that includes the word *reality* is scientific, since no scientist in his right mind would attempt to define something as slippery as reality, much less study it. Still, if the question is broken down into parts, maybe something scientific can be made of it.

For example, it can be conjectured that people drink for one of two reasons (leaving aside drinking that is strictly a social act): (1) They drink to feel pleasure, or (2) they drink to escape pain (the latter being Roueché's explanation).

Some argue that pleasure and release from pain are identical. Urination, the theory goes, is pleasurable because the urge to urinate is unpleasant; the stronger the urge, the more pleasant its abolishment. This makes sense and people are quick to say, "Yes, that's the way it is."

But then examples are cited that don't fit. Is the sexual act pleasurable because it abolishes sexual desire? With orgasm usually comes an abrupt termination of sexual desire, but does this entirely explain orgastic pleasure? Most people, from personal experience, would doubt it. Are chocolate bars, for people who like chocolate bars, pleasurable to eat *only* when hungry? Chocolate addicts know better. To return to the subject, is alcohol enjoyable only when a person is frustrated, insecure, lonely, miserable? While acknowledging alcohol's power to succor, most drinkers on reflection will say alcohol does more than that—it makes them feel good even when they are *not* feeling bad.

Of course, maybe drinkers feel bad without knowing it, but the argument can go on forever, since it is a philosophical argument and not a scientific one. However, perhaps something scientific can be made of it.

Let's assume that alcohol affects the brain in two ways: It has *sedative* properties, dulling sensation and responsiveness, and it produces *euphoria.* Let's further assume that these effects are independent and are related to different structures or chemical systems in the brain. Have we advanced much further? Can this be studied? It can, and has been, although still in a preliminary way.

Certain centers in the brain, for example, appear to regulate sleep. Stimulating some centers electrically or chemically puts animals to sleep; destroying other centers keeps them awake. Another part of the nervous system functions as an "activating" system and is essential in maintaining vigilance and wakefulness. Alcohol may affect these areas selectively—there is some evidence that it does—which would locate, if not explain, its sedative effects.

In the 1950s a new term was introduced by experimental psychologists: the pleasure center. It came from the discovery that cats and rats will press a bar repeatedly if the pressing delivers a tiny charge of electrical current to certain brain centers. Since there was no better explanation for this seemingly pointless behavior, the observers (recalling the pointless behavior people sometimes engage in) concluded that the animals were pressing purely for pleasure and therefore the centers must be "pleasure centers."

Well, pleasure centers have been controversial and are not as popular now as they were in the 1950s, but nobody has come up with a better idea why animals will stimulate themselves so steadily; and, if there are pleasure centers, maybe alcohol has some kind of affinity for them. More work, as the saying goes, is needed.

Finally, a lot of time and money has been spent in recent years studying chemicals in the brain called amines. A theory has emerged from these studies which, like pleasure centers, is not as fashionable as it once was but still remains pretty popular. The theory holds that elation is related to high concentrations of certain amines and depression is related to low concentrations. This has been an attractive theory for researchers who study alcohol. Alcohol, it seems, raises or lowers the concentration of these amines—the studies are inconsistent—and *may* be the chemical basis for the mood-raising, mood-lowering effect of alcohol described in chapter 2. Such speculation may seem premature, but at least it leads to testable hypotheses, which is what science is about.

QUESTION 2: WHY DRINK A LOT?

Behavior, it is widely believed, is motivated; that is, people do not do things without a reason. If some people drink more than others, there obviously is a reason. (That is, if you believe in cause and effect; if you do not, this chapter is not for you.)

Probably the simplest reason some people drink more than others is that alcohol does more for them. It gives them more pleasure, spares them more pain, or both. Why?

As discussed earlier, alcohol affects individuals differently and for reasons that are almost surely biological. Orientals tend to flush when they drink; Occidentals tend not to (see the end of chapter 5). Some people can't drink much because a little alcohol makes them sleepy or sick. Also, alcohol affects moods differently. It makes some people higher than others. The variable euphoric effects of alcohol may be as innate, as biological, as flushing or vomiting. If so, it may explain partly why some people drink more than others—they like it more.

Or—back to Roueché—maybe they dislike reality more. Their need to escape is greater. Maybe reality is rougher on them than on others—they have more heartaches and toothaches, bunions and bad times—or they are born more sensitive (or become that way) to slings and arrows that others can tolerate without the armor of alcohol. Or maybe, just as drink makes some people happier than others, it sedates some more than others. Or perhaps all of these factors are operating, in various combinations, to produce that familiar figure, the heavy drinker.

These are a lot of maybes and they have not led us very far. Knowing more about why people drink will probably help explain why some people drink more than others. Those scientists who are trying to make sense out of sleep centers, pleasure centers, and why amines go up and down may yet provide the key.

QUESTION 2 (CONTINUED): WHY TOLERANCE?

Another reason some people drink more than others is that they *can* drink more: they have tolerance. More accurately, perhaps, they lack intolerance. They don't get sick, dizzy, or fall asleep when drinking; their hangovers are bearable. Is tolerance inborn, acquired, or both? Here at last is a question that can be studied:

Select teenagers before they have become accustomed to drinking much, if at all. Give them alcohol. Measure the efects. Follow them into adulthood. If inborn tolerance contributes to heavy drinking, the teenagers who could drink the most with the least effect will—the hypothesis goes—drink the most as adults.

In the past few years, such studies indeed have been done. They indicate that some teenagers can drink more than others without showing the effects. These teenagers, it happens, usually have alcoholics in the family. Follow-up studies are needed to determine whether the teenagers with the greatest tolerance are those who develop drinking problems. At any rate, the evidence now is fairly strong that tolerance for alcohol (the ability to drink a lot) is partly innate. More about follow-up studies a little later.

Some may object to giving alcohol to teenagers, and for sound ethical reasons. As with all human research, the potential benefit from the study must be weighed against the possible risks. The risk of such a study is that giving alcohol to young people may in some way harm them. What benefit is there in knowing whether tolerance is innate or not? In fact, the knowledge might lead to something useful. If tolerance is innate and if it increases the chance of a person becoming a heavy drinker or alcoholic, understanding the chemical basis for tolerance might lead to ways of altering tolerance. This in turn might result in a specific treatment for alcoholism, a possibility discussed in the "Let's Be Practical" section below.

QUESTION 3: WHY ALCOHOLISM?

Drinkers can be divided into moderate drinkers, heavy drinkers, problem drinkers, and alcoholics. It has been proposed that heavy drinkers drink for the same reasons that other people drink, but that alcohol does more for them and so they drink more. Is this true of alcoholics? Does alcohol do even more for them, and hence they drink even more?

Or does alcohol affect alcoholics in a special way—in a way that is incomprehensible to normal drinkers and sets the alcoholic apart, perhaps from birth, as a person destined to drink abnormally?

To many normal drinkers, alcoholism *is* incomprehensible. It's easy to understand why people drink: they enjoy it. It's not easy to understand why people drink themselves to death. Nonalcoholics have hangovers, swear they'll never do it again, and eventually do it again anyway. Even this is understandable: memory is short, the will is weak, and so forth. But when a person wakes in the morning after a drinking bout, vomits, has a drink, vomits, has a second drink, vomits, and prays the third will stay down—this is not so easy to understand. What maintains behavior that results in such suffering? Drinking may be fun, but not *that* much fun.

In fact, there is evidence that alcoholics may differ from heavy drinkers in more than degree; there may indeed be something special in their response to alcohol. In the Danish study discussed in chapter 5, sons of alcoholics who were raised apart from their alcoholic parents more often became alcoholic than did sons of nonalcoholic, but they were no more likely to become heavy drinkers. It appeared that alcoholism was inheritable, but that heavy drinking was not. There is other evidence to support this. For example, alcoholism that begins in early life seems to have been more influenced by heredity than alcoholism that begins later. However, more evidence is needed. How does one answer such difficult questions as, "Does alcohol affect alcoholics differently than it affects other people?" or, conversely, "Are alcoholics people with specific problems who turn to alcohol because it is available, as they would turn to another drug that gave them comparable relief if it were available?"

As mentioned earlier, one approach is to conduct a "high-risk" study. To recapitulate: Children of alcoholics have a higher risk of becoming alcoholic than do children of nonalcoholics. Study these children as children. Obtain all kinds of information about them—physical, psychological, social—starting in infancy or even before (using pregnancy records). Follow them through the years into adulthood. Some almost certainly will become alcoholics. Why *them*? From the mass of information by then available about them it should be possible to point to some factors.

To know something about these factors could make prediction possible. And if you can predict, possibly you can prevent. And prevention is the best treatment of all.

As noted, high-risk studies are being done. Marc Schuckit at the University of California in San Diego pioneered these studies and has produced numerous papers reporting his results. Roger Myers and associates are conducting other studies at the University of Connecticut, and Henri Begleiter and associates are conducting research at Downstate Medical Center of New York University. The author is involved in a high-risk study in Denmark. Initial results from these studies have been interesting (see chapter 4), but insufficient time has passed to permit the follow-up studies that will provide the real payoff. This involves matching individuals who develop drinking problems with information obtained about them before they began drinking. The follow-up studies will begin in a few years.

SWITCHES

Some people say that alcoholism should be called alcohol*isms:* There is not one alcoholism but many. The evidence that there might be two forms of alcoholism—familial and nonfamilial, or Type 2 and Type 1—has been presented elsewhere in the book. The evidence, so far, is inconclusive.

Even proponents of the one-alcoholism theory hold that the cause of alcoholism is multifactorial. *Multifactorial* is a mathematical-sounding word usually uttered in tones which suggest that our ignorance about alcoholism may be total. In fact, the cause of alcoholism is not known, meaning that it could just as easily be unifactorial as multifactorial. We probably should admit this and not disguise our ignorance with fancy language.

Perhaps alcoholism is called multifactorial because that seems diplomatic. Meetings on alcoholism are attended by people from a variety of specialties: medicine, psychiatry, psychology, social work, among others. Everyone wants a piece of the action. When alcoholism is attributed to multiple factors—biologic, sociologic, psychologic—everyone feels useful. There is an unspoken agreement that, because experts from diverse backgrounds study alcoholism, alcoholism must have diverse origins.

In one sense this is obviously true. Genes give us the enzymes to metabolize alcohol; society gives us the alcohol to metabolize; and our psyches respond in wondrous ways to these combined gifts. Nevertheless, beyond this obvious level, the evidence for multiple causes of alcoholism is no more or less convincing than the evidence for a single cause. The cause of alcoholism is simply unknown.

Alcoholism may actually involve a single cause—a single chemical "switch." If someone ever finds the switch, the next step will be to learn how to turn it off. This should be relatively easy, and has been done in other illnesses. Lewis Thomas, our best writer in medicine, says the same thing about cancer. When people interested in cancer get together at meetings, they also use terms like multifactorial. But Thomas believes cancer may also involve a single switch:

> The record of the past half century has established, I think, two general principles about human disease. First, it is necessary to know a great deal about underlying mechanisms before one can really act effectively. . . .
>
> Second, for every disease there is a single key mechanism that dominates all others. . . . This generalization is harder to prove . . . but I believe that the record thus far tends to support it. The most complicated, multicellular, multitissue, and multiorgan diseases I know of are tertiary syphilis, chronic tuberculosis, and pernicious anemia. . . . Before they came under scientific appraisal each was thought to be what we now call a "multifactorial" disease, far too complex to allow for any single causative mechanism. And yet, when all the necessary facts were in, it was clear that by simply switching off one thing—the spirochete, the tubercle bacillus, or a single vitamin deficiency—the whole array of disordered and seemingly

> unrelated pathologic mechanisms could be switched off, at once. . . .
>
> I believe that a prospect something like this is the likelihood for the future of medicine. I have no doubt that there will turn out to be dozens of separate influences that can launch cancer . . . but I think there will turn out to be a single switch at the center of things. . . . I think that schizophrenia will turn out to be a neurochemical disorder, with some central, single chemical event gone wrong. I think there is a single causative agent responsible for rheumatoid arthritis. . . . I think that the central vascular abnormalities that launch coronary occlusion and stroke have not yet been glimpsed, but they are there, waiting to be switched on or off.

Thomas did not include alcoholism in his list of potential single-switch disorders, but might have, had he attended a few multidisciplinary conferences on alcoholism.

Fifty years ago, a meeting was held on the topic of tuberculosis. This was the era of the "TB methodologist." The discussions revolved around topics such as sending patients to the mountains versus the desert versus the seashore; whether bathing in salt water was better than not bathing in salt water; and how you could tell whether such things made a difference. Evidence was presented that going to the mountains led to improvement in 15 percent of patients, whereupon a discussant pointed out that people who went to the mountains usually had more money and were better nourished than people who did not go to the mountains, which might explain the improvement rate. The term *confounding variable* was as popular at that meeting on tuberculosis fifty years ago as it is in current conferences on alcoholism.

A few years later, the "switch" was found in tuberculosis. You *had* to have the bacillus to get the disease. You could have the bacillus and *not* have the disease, but you could not get the disease without the bacillus. If you got rid of the bug, you got rid of the disease. Since that discovery there has not been another meeting on the methodologic aspects of studying tuberculosis.

There are two possible switches that may turn on alcoholism. These are based on the fact that, in the world of switches, there are two broad possibilities: one produces too little of something or one produces too much of something.

I have a friend who illustrates the first possibility, the deficiency hypothesis. His wife says he was born "two martinis below par." Ordinarily, he is a surly fellow who comes home, hides behind a newspaper, and will not communicate with anyone. But give him two martinis and he becomes a jolly, gregarious man who not only tells his wife what happened at the office but listens to what happened at home. He never has more than two martinis. He is not an alcoholic. There is an element of dependency, but it is mainly his wife's dependency on his having two martinis. Not a problem the various alcoholism groups would pounce on.

Is the change chemical? It seems likely. What chemicals? It is too early to tell; the sensitive tools needed to measure such things have only been around a short time, and the government only spends $2 on research for each alcoholic in the country, compared to $250 for each cancer patient, according to data provided by William F. Mayer, former head of the federal mental health, drug and alcohol agencies. Conferences, prevention centers, and the social sciences eat up much of the $2, leaving a pittance for chemistry. I do not object to conferences, prevention centers, or the social sciences, but I am concerned about the pittance. Placed properly, money might enable us to find a chemical basis for the following hypothetical events:

Alcoholics, or some alcoholics, lack the gene(s) for optimal production of Substance H (H for happiness). They are born, so to speak, unhappy. Sometime in their teens they discover that having two or three beers or drinks makes them happy. Someone once said that you never feel better than when you start feeling good after feeling bad, and, for some alcoholics—those who start life feeling bad—drinking alcohol feels so good it becomes habit-forming. But all is not bliss. Ten minutes or so after the H-deficient person has a drink, he wants another. He is feeling unhappy again—unhappier than before he had the drink. The unhappiness now has a special quality sometimes called craving. A drink relieves this new unhappiness and even restores the original happiness, but just briefly. Another drink, and another, is needed to overcome the unhappy feeling produced by the same drug that produced the happy feeling—alcohol. "It lifts you up and it lets you down," as the saying goes, and some people are lifted up and let down more than others, probably because of genes.

Substance H (if it exists) has not been identified. Let's assume, for the sake of speculation, that it is the brain neurotransmitter serotonin. Perhaps the prealcoholic has too little serotonin. Brain cells contain limited amounts of serotonin and, after a short time, they run out of the neurotransmitter. Alcohol releases serotonin from brain cells and brings the serotonin level to normal and even above normal, making the act of drinking highly reinforcing. Now the prealcoholic who had low levels of serotonin to begin with has even lower levels, and because drinking, for him, is so strongly reinforced and habit-forming, he begins to wage a desperate battle to overcome his serotonin deficiency by drinking more and more alcohol. But it is a battle he will inevitably lose; his cells cannot keep up with the demand for serotonin. As a result, he feels considerable stress and strain. Stimulant chemicals like epinephrine pitch in to help, but ultimately make matters worse by leaving the drinker in a hyperstimulated state when he stops drinking (as he eventually must).

Reserpine has this release-and-deplete effect on serotonin in brain cells; what was described above could be called "the reserpine model of alcohol addiction." The reserpine model works for many other chemicals, including prostaglandins, according to some studies. It is unlikely that Substance H is really serotonin, but with sufficient prodding of chemists and with taxpayer dollars, someday the right substance that fits this or another model may be found.

I mentioned a second chemical possibility, the overproduction hypothesis. This argument assumes that alcohol produces substances in the brain that resemble morphine. Conceivably, alcohol produces more morphine-like substances in some brains than in others—again because of genes. The prealcoholic is the person whose brain overproduces morphine when he drinks alcohol—or overproduces *something* that makes drinking unusually rewarding, habit-forming, addictive.

Chemical mediators of addiction (if they exist) may not be like morphine at all, but the overproduction model is a good one to pursue. In fact, if I were allocating dollars, I would allocate as much as could profitably be spent on the continued study of morphine-like chemicals produced by alcohol in the brain. Whether it is a blind alley would be learned fairly quickly and, meanwhile, talented people would be encouraged to study other potential causes of alcoholism.

Perhaps I'm naive about what chemistry has to offer. A magic bullet for alcoholism may be years away—or may not exist. In the meantime, research on psychosocial factors in alcoholism is vitally important. We need to know what helps alcoholics. We need predictors. Each time a reliable predictor is discovered, we move a step closer to causality. The tubercle bacillus would have been just another bug seen through a microscope if clinical research had not first assembled a disease to go with it.

Meanwhile . . .

LET'S BE PRACTICAL

What can be done to help alcoholics now or in the near future, without waiting years for the outcome of a high-risk study or for breakthroughs in brain chemistry? Doctors often can help people without knowing what is wrong with them. This is fortunate, since they rarely know what is wrong, or at least the precise cause of what is wrong. This helping-without-knowing-why is called empirical medicine, which even in our vast collective ignorance helps many of us stay healthy.

Here are some things that might help:

1. *Develop a better Antabuse.* This is a drug that makes people sick when they drink but usually has no effect if they

do not drink. It has been helpful in the treatment of some alcoholics (see chapter 8) but has two drawbacks: (1) people find reasons not to take it, and (2) combined with alcohol it can produce a serious and sometimes fatal reaction.

A superior Antabuse would do the following: its effects would last at least several weeks so that a single decision to use the drug would provide long-lasting protection, and it would have milder effects if combined with alcohol. Ideally, the latter effects would be proportionate to the amount of alcohol consumed. A person taking the drug could go to a ball game, for example, and have a couple of beers with no adverse effects. But if he drank much more he would begin to feel queasy, dizzy, or have some other unpleasant reaction. In other words, his response to alcohol would resemble that of millions of other people who can drink a certain amount and enjoy it but develop distressing symptoms if they drink more. In a sense, the drug would convert alcoholics' high tolerance to alcohol to low-to-moderate tolerance.

The drug might have an advantage other than deterring individuals from excessive drinking. It might permit them to learn a new pattern of drinking that would persist even without the drug. With repeated experiences of feeling poorly after drinking a certain amount of alcohol, conditioned responses might be established that would deter the individual from drinking that much. Normal drinking would become, in the literal Pavlovian sense, habitual. In attaining the goal of normal drinking—in this boozy society perhaps the ultimate goal—some form of psychological treatment might be helpful. Drug therapy is perfectly compatible with other forms of treatment, including psychoanalysis, transcendental meditation, and Skinnerian conditioning. Indeed, drug therapy sometimes makes these approaches possible when they may not be otherwise.

2. *Develop a sobering-up pill.* For years people have looked for ways to speed up the elimination of alcohol from the body. The theory was that if alcohol went away faster, hangover effects would be less unpleasant and, to the extent that people drink to cure hangover, this particular motivation to drink

would be less powerful. In fact, a case could be made that the hangover would be worse if the alcohol disappeared faster. Still, speeding up its elimination would serve one purpose: it would shorten the period of intoxication.

Many things have been tried—insulin, caffeine, exercise—but only one has worked. Fruit sugar (fructose) in large doses definitely speeds the elimination of alcohol. Unfortunately, the dose required is so large it is sickening, and most people prefer to remain drunk. It is not known how fructose speeds the elimination of alcohol, but hard investigation may lead to a substance that would hasten the removal of alcohol from the body without ill effects.

The pursuit, at any rate, would seem worthwhile. Emergency-room doctors could give the drug for alcoholic coma that threatens to be fatal. Bartenders could give it to customers; liquor-serving hosts could give it to guests before they drive home. One can imagine a bottle of pills in every glove compartment. Alcohol intoxication causes thousands of deaths every year on the highway. A rapid-acting, safe, effective sobering up pill could reduce this figure radically.

3. *Develop an "anticraving" pill.* This substance would decrease the need to drink. Kenneth Blum of the University of Texas is working on such a pill, but at this writing, studies had not yet demonstrated its usefulness. Animal data indicate that natural substances that increase endorphin levels (see chapter 1) might block craving, but the development of such a pill for humans may still be a bit futuristic.

4. *Vitaminize.* Some of the physical ills that people suffer from heavy drinking are caused by malnutrition. Specifically, they are caused by a deficiency of B vitamins. Being water soluble, these vitamins could be added to alcoholic beverages. People do not complain about enriched bread and probably would not complain about enriched bourbon. It might all but eliminate the rare but catastrophic brain diseases associated with alcoholism.

5. *Work on the liver.* Much has been learned in recent years about the effect of alcohol on the liver, about the intricate interactions of alcohol, fat, protein, and carbohydrates. Con-

tinued large-scale effort in this area might, in a relatively short time, result in drugs that would cure or dietary changes that would prevent the fairly common types of alcohol-related liver disease. It is possible that alcoholic cirrhosis will be curable before alcoholism is.

6. *Educate.* This seems easy, but it isn't. First, you have to know what to tell people, but experts can't agree among themselves, so even this is difficult. Then you have to get people to pay attention. People can spot propaganda a mile away. The preachy kind of propaganda is particularly offensive to many. Credibility is all-important. Once young people decided they couldn't believe anything adults told them about marijuana, they stopped listening, and even if marijuana were conclusively shown to cause feeblemindedness, gum disease, and cancer, they might pay no heed.

Education, indeed, may produce the results it tries to discourage. Some of the early antidrug films made drugs appear so attractive that people were probably tempted to use them when they might not have done so otherwise. At the very least, the films publicized drugs' existence.

Nevertheless, who can argue against education if the information is reliable? Certain diseases do run in families, and families so involved should at least be aware of it. Children of alcoholics should at least know that their chance of becoming alcoholic is greater than most people's.

STUDIES THAT NEED REPEATING

Alcoholism research has been plagued by erraticism. Here are some reasons:

Few investigators make a career of alcoholism research. Like alcoholism itself, alcoholism research has tended to be unpopular in the medical and scientific communities, probably because people think of it in terms of morals rather than science (and besides, drunks are dirty, smelly, etc.). Over the years most studies about alcohol and alcoholism have been done by people who tossed off the study between other work, so to speak, because suddenly some alcohol research money dropped in their lap or they happened to be working with someone interested in the subject or they needed a Ph.D. thesis and there was a captive audience of alcoholics at the local VA. Perhaps some went into it to find an answer to their own alcohol problem, but probably not many.

What happened, with so many people dashing in and out of the field, leaving a paper or two behind, and then going on to other interests, was that the wheel kept being invented and there weren't many people around with a long perspective, or even people who knew the literature. ("Every twenty years people repeat the same studies," someone said, "because they never read papers before twenty years ago.")

Alcoholism research has also been plagued by erratic funding. Before the late 1960s, foundations and the federal government spent almost nothing on alcohol research. The little work that was done consisted of spin-offs from other studies or the straying of some physiologist down an unfamiliar byway. Even after the federal government began supporting research, the money ebbed and flowed. It mainly ebbed. The money was a vacuum that sucked in people of talent, who then found, after a success or two, that they couldn't get sustained support. Alcoholism research centers were fashionable for a time, then lost favor, and now are fashionable again. Yet their funding is still uncertain. Investigators in cancer or AIDS can count on support, but not in matters as influenced by fad and fashion as drug and alcohol research.

Despite the haphazard approach, findings have been reported from time to time that, if true, might open doors to the mystery of why most people drink and why some become alcoholic. Unfortunately, the sad situation is that hardly anyone attempts to repeat someone else's work, however important the findings are. When they do, the attempts to repeat the work are often flawed by a malicious motive to show that the other person was wrong or by a failure to repeat the work exactly as it had been done originally. This failure to repeat work exactly has contributed to the many contradictory reports in the literature. The Institute of Alcohol Abuse and

Alcoholism would be well advised to have a Department of Replication staffed by solid scientists who do nothing but attempt to repeat other people's important findings, using exactly the same methods used by the original investigators.

Another casualty of the high drop-out rate in alcoholism research is the abandonment of longitudinal studies. Perhaps a dozen groups of investigators have recently formed cohorts of prealcoholic sons of alcoholics and collected data about them with the intent of following the subjects into middle life. If they are not followed, the study is wasted. In dank governmental and university building basements around the country are cardboard boxes full of files and computer tapes from longitudinal studies that never went beyond the initial phase. Investigators went into something else, or died, or lost funding, or lost interest.

Such waste should not be allowed. Some questions cannot be answered without longitudinal studies, and the answers are needed. There should be some mechanism in the government that assures longitudinal studies will truly be longitudinal. Funding should be assured for thirty or forty years, if necessary. There should be a way to assure continuity, even if it means recruiting investigators to replace those who have fallen by the wayside.

Meanwhile, here are some studies that need repeating:

1. The only genetic marker study that has been successfully replicated was the study showing that alcoholics more often were nonsecretors of ABH substance than were nonalcoholics, particularly those with blood group A (see page 132). It needs repeating. Reliable markers would help predict individuals at risk for alcoholism.

2. The P300 study described in chapter 3 needs additional repeating. It is too important to ignore.

3. A recent study found that some individuals show more state-dependent effects from alcohol than others and that these individuals are particularly susceptible to blackouts (see page 34). They also may be more susceptible to alcoholism. The study needs repeating.

4. Alcohol has been reported to have roller coaster effects on serotonin, endorphins, and prostaglandins. The addictive cycle (chapter 2) also resembles a roller coaster, and the underlying chemistry needs explaining.

5. TIQs should be pursued, despite deep skepticism (chapter 1).

6. No alcohol receptors have been found in the brain, but receptors for Valium-type tranquilizers have been discovered and alcohol seems to influence them. Chronic alcohol use may reduce the number of receptors, or maybe some people are born with fewer receptors and alcoholism comes from an attempt to correct the deficiency. Fewer receptors means more anxiety—and a subgroup of alcoholics are anxiety-prone. This is worth pursuing.

7. A study reported that sons of alcoholics generated more alpha "serenity" waves on the EEG after drinking than do controls (see page 94). This study is almost too good to be true—it seems to indicate an inheritable, and powerful, reason to drink—and should be pursued.

8. If alcoholism reflects a genetic defect, how can it be pinned down? Gene mapping is one approach. In 1978, scientists introduced a method for gene mapping called restriction fragment-length polymorphism (RFLP). Without explaining what this mouthful means, the importance of RFLP has been universally acknowledged. In 1983 the gene for Huntington's chorea, a devastating neurological illness, was tracked to a small region of chromosome 4. It was a landmark study. It showed that genes for an illness could be located on chromosomes even when there was no inkling of where the genes might be. Molecular genetics is a wide-open field, with new technologies appearing regularly. The entire human genome—the total set of genetic blueprints contained in DNA—may someday be as readable as the works of Shakespeare.

Is there a word or phrase or sentence in the genome that makes some people more susceptible to alcoholism than others? Of course, this is not known, but the technology is now there to find out.

CHAPTER EIGHT

Treatment

> One thing about alcohol: it works. It may destroy a man's career, ruin his marriage, turn him into a zombie unconscious in a hallway—but it works. On short term, it works much faster than a psychiatrist or a priest or the love of a husband or a wife. Those things . . . they all take time. They must be developed. . . . But alcohol is always ready to go to work at once. Ten minutes, half an hour, the little formless fears are gone or turned into harmless amusement. But they come back. Oh yes, and they bring reinforcements.
>
> —Charles Orson Gorham

Alcoholism treatment has become a major industry with important ramifications. But is alcoholism treatment effective? Sometimes people recover from an illness without professional help. This is called spontaneous remission; "spontaneous" means the remission cannot be explained. It happens in almost every illness, including alcoholism. Before treatment can be judged effective, it must be shown to be superior to no treatment.

Usually, studies comparing treated and untreated groups are needed in order to show effectiveness. Some treatments are so effective, however, that they obviate such studies. (Using penicillin to treat pneumonia is an example.) But it is a mistake to judge the effectiveness of a treatment by what happens to one, two, or a small group of patients. Even terminal cancer patients sometimes recover "spontaneously," and the history of medicine is a graveyard for treatments that were worthless but flourished because people tend to get over things anyhow.

There is no penicillin for alcoholism. Studies are needed. Some of the treatments discussed in this chapter have never been studied; others have been studied but not well. It may be a slight exaggeration, but only slight, to say that no study exists which proves beyond question that one treatment for alcoholism is superior to another treatment or to no treatment.

Nevertheless, alcoholics seek help and people try to help them. Uncertainty about the results of treatment has not and should not discourage this effort. Who is providing treatment? What treatments are available?

Alcoholism treatment has evolved into a major industry in the past few years. An estimated 15 percent of the national health bill is spent on alcoholism and alcohol-related problems, such as cirrhosis.

There has been a rapid proliferation of for-profit treatment facilities, many of them not located at hospitals, some charging up to $10,000 a week. The twenty-eight-day inpatient stay has become the cornerstone of alcoholism treatment in the United States. General hospitals have found that alcoholism rehab is an excellent way to fill surplus beds. Advertising for the programs has become blatant and sometimes sleazy. There are even services for family members of alcoholics—beds offered for the "disease of co-dependency," referring to the distress of living with an alcoholic.

The twenty-eight-day programs offer a mishmash of services, ranging from films, group therapy, relaxation training, vocational counseling, family therapy, and Alcoholics Anonymous sessions to the use of medications, particularly Antabuse. According to Enoch Gordis, director of the National Institute on Alcohol Abuse and Alcoholism, this has resulted in a "haphazard mixture of largely unvalidated approaches." Assessing treatment, Gordis says, is extremely difficult. More than 50 percent of patients drop out of these programs. Little is known about the natural history of the untreated alcoholic. Self-reports of drinking behavior are notoriously unreliable.

Short-term follow-ups may be misleading because someone who resumes drinking may quit later.

One thing seems clear: As a whole, inpatient programs are no more successful than outpatient programs. For certain patients, inpatient treatment may be best, but it is difficult to know who these patients are.

In a 1986 paper, William R. Miller of the University of New Mexico reviewed twenty-six treatment studies and concluded that they have "consistently shown no overall advantage for inpatient over outpatient settings, for longer over shorter inpatient programs, or for more intensive over less intensive interventions."

"Intensive" is synonymous with costly. Helen Annis of the Addiction Research Foundation of Toronto arrived at the same conclusion: Lengthy inpatient programs provide no better success than brief hospitalization, day treatment programs, or outpatient programs. Annis says that "one of the disappointments over the past ten years" has been that neither intensity nor type of treatment is a good predictor of success. As the Institute of Medicine put it in a 1980 report: "The best predictor is the patient."

What does this mean? It means that regardless of treatment one can predict with fair success those patients who will do well. They are people with jobs, stable marriages, minimal psychiatric disturbances, no history of past treatment failures, and minimal involvement with other drugs. Severity of alcohol abuse is neither a predictor of treatment success nor a failure.

The lack of relationship between "intensity" of treatment and success is graphically illustrated by a 1967 study by Griffith Edwards and his colleagues at the Maudsley Hospital in London. They compared patients who received several weeks of inpatient treatment with patients who had a single session with a counselor and a monthly checkup. At the end of a year both groups were doing equally well—or equally poorly. The patients had been randomly assigned to the two groups.

Who does the actual treatments? Treators include clergy, social workers, psychologists, psychiatrists, other physicians, and people called alcoholism counselors, many of whom are recovering alcoholics. Most large cities have a branch of the National Council on Alcoholism listed in the phone book. Among other things this useful organization maintains a list of alcoholism-treatment services.

What kind of treatments are offered? They include psychotherapy, behavior therapy, medications, and referral to Alcoholics Anonymous.

PSYCHOTHERAPY

There are many schools of psychotherapy, but they all have one thing in common: They involve two or more persons talking to each other, and one of the discussants is supposed to know more than the other about what is going on. He knows more because he has gone to school and received a degree or has had some other special experience that makes him an authority on the subject, such as having been in treatment himself or having himself been a victim of the illness being treated.

"Talking it out" or "getting it off your chest" has long been valued as a means of relieving certain forms of distress, such as a smoldering resentment or feelings of bereavement. The amount of relief thus attained, however, is often exaggerated, and sometimes, rather than feeling better for having talked about their problems, people feel worse; talking about the problems reminds them of them. At any rate, psychotherapists tend to attribute the success they believe they obtain to unique features of their particular school of psychotherapy.

Each school has a doctrine. Each doctrine, almost without exception, has a founder, a great man whose theories and writings are highly esteemed by his followers, so that schools of psychotherapy inevitably tend to resemble religions. The therapist is a votary and the patient a supplicant hoping to overcome his problems by *understanding* them. In psychotherapy, understanding has roughly the same role as grace in religion. There is an assumption in psychotherapy which is sometimes denied but is almost always present—that understanding the nature and origins of a particular problem helps a person overcome the problem.

This has come to seem almost self-evident, but it is not clear why. A patient with cancer presumably would benefit little from understanding the cause of his condition. What we call understanding, in fact, is simply the point where curiosity rests. There is no area of inquiry where you cannot go a layer deeper if you are able to and so desire. Electricity is understandable as a "flow of electrons" but what is an electron? According to the dictionary, an electron is an "elementary particle consisting of a charge of negative electricity," which shows how understanding is frequently an exercise in circular reasoning. In psychotherapy understanding occurs when the patient agrees with the therapist about what is wrong with him. When the patient's theories about himself coincide with the therapist's theories, he is supposed to improve. Does he?

There is hardly any *scientific* evidence that psychotherapy for alcoholism or any other condition helps anyone. There have been a number of studies, but few have met even the minimal requirements for a scientific study. Yet thousands of people make their living giving psychotherapy and millions have received psychotherapy, many of them feeling they have benefited from it. They indeed may have benefited from it; after all, there is no evidence that psychotherapy does not work. But the burden of proof, as always in such matters, is on the proponents.

The schools of psychotherapy are too numerous to go into here. Most are identified by the name of their founder. Much of what is called psychotherapy can be subsumed under two headings: psychodynamic psychotherapy and transactional analysis.

Psychodynamic therapy is generally of two types—the classical type and the nonclassical or watered-down type. Both ultimately derive from the doctrines of Freud and his followers. The classicists attribute mental illness to unconscious conflicts that originate in early childhood. Since drinking is an oral activity, alcoholism is assumed to arise from oral conflicts.

Small babies are more oral than anyone, since practically their whole existence revolves around sucking a nipple. If the sucking does not go well, conflicts develop which, unless resolved in psychoanalysis, result in an "oral-dependent" personality (dependent because tiny infants are exceedingly dependent on mothers). Alcoholics are said to have oral-dependent personalities. They not only drink too much, but if you go to AA meetings you find they also smoke a lot and drink gallons of coffee, as well as talk a blue streak when given a chance. Alcoholics can overcome their oral dependency, but it takes a long time. They must first have "insight" about their orality and, second, "work out" their oral conflicts by being dependent for extended periods on their therapists. They are encouraged to believe this, despite the widely held view among psychoanalysts that alcoholism is exceedingly difficult to treat. Freud himself held that addictions were hard to treat because, at bottom, they were so pleasurable.

Another psychoanalytic explanation for alcoholism is that alcoholics are latent homosexuals. The reasoning is that since both homosexuality and drinking involve oral activity, both have origins in oral conflicts. It explains, too, why male alcoholics seek out the masculine camaraderie of barrooms; they can be with men while simultaneously denying their homosexual tendencies by engaging in heavy drinking, an activity associated with masculinity. Others hold that alcoholism is a form of self-destruction (which it obviously is) and has the same roots as depression. People have angry feelings toward others, cannot express them, and therefore become angry at themselves. Self-hatred is subjectively experienced as depression or leads to self-destructive acts such as alcoholism. Therapy consists of helping a person recognize his unconscious drives and motives, which results in a happier and more mature person who does not drink as much.

Many therapists who are not technically psychoanalysts nevertheless use concepts derived from Freudian theory. Any therapist who uses such terms as "ego defense mechanism" or "acting out" views abnormal behavior, whether he realizes it or not, as a product of psychological conflict. This, at bottom, is what psychodynamic means. Although everyone has experienced psychological conflict, the idea that mental illness arises from conflict remains speculative.

A formerly popular form of psychotherapy now somewhat in decline, was based on the theories of Eric Berne and called transactional analysis. Dr. Berne was an amusing writer whose ideas have much in common with Freud's. Instead of id, ego, and superego, he substitutes the terms child, adult, and parent. These three dramatis personae of mental life are constantly feuding, just as members of a family quarrel, and the quarrels sometimes take the form of abnormal behavior. All people play games, and "sick" people play games calculated to make them losers. The alcoholic punishes other people by punishing himself and is the loser in the end. Transactional analysis, or TA, makes alcoholics aware of the games they play and encourages them to find other games that are less destructive.

Transactional analysis lends itself somewhat more to group therapy than does psychoanalysis or psychodynamic psychotherapy. It is often said that group therapy helps alcoholics more than individual therapy does, but there is no evidence for this. Group therapy does have one advantage: it takes less of the therapist's time.

Anyone reading this who believes the theories of Freud, Jung, Adler, Sullivan, Berne, or any of the other founders of psychotherapeutic movements to be correct will no doubt be offended by the above or at least take strong issue with what has been said. This is unfortunate, but there is no avoiding it; if there is no acceptable evidence that something works, it would be unfair to the reader to pretend that there is. Single-case reports that patients have improved while receiving a particular therapy do not constitute acceptable evidence. As noted earlier, even terminal cancer patients sometimes recover.

To show that a particular treatment is useful, three questions must be asked: Would the patient have recovered without any treatment? Would he have have done as well or better with another treatment? Is the improvement related to nonspecific aspects of a treatment? "Nonspecific aspects" include, in Peter Medawar's words, the "assurance of a regular sympathetic hearing, the feeling that somebody is taking his condition seriously, the discovery that others are in the same predicament, the comfort of learning that his condition is explicable (which does not depend on the explanation being the right one)." These factors are common to most forms of psychological treatment, and the good they do cannot be credited to any one treatment in particular.

ALCOHOLICS ANONYMOUS

Since its creation in the 1930s by two alcoholics, this organization has grown into a worldwide network of self-help services for alcoholics and their families. It has many attractive features, including three of Medawar's four common denominators of psychotherapy (and at no cost): the assurance of a regular sympathetic hearing, the feeling that somebody is taking one's condition seriously, the discovery that others are in the same predicament. Unlike most talking therapies, however, AA expends little effort in trying to explain *why* anyone is alcoholic. The term "allergy" is sometimes used, but usually properly bracketed in quotation marks (alcoholism does not of course resemble the conventional allergies at all).

There is an old idea that alcoholics must become religious in order to stop drinking, and it is true that Alcoholics Anonymous has certain similarities to a religion and that some of its members have been "converted" to AA in the same way they would be to other religions. Its "twelve steps," for example, have a definite religious flavor, emphasizing a reliance on God, the need for forgiveness, and caring for others.

Nevertheless, to the extent that it is a religion, AA is one of the least doctrinaire and authoritarian religions imaginable. Atheists can belong to AA as comfortably as believers. There is no formal doctrine and no insistence that anyone accept a particular explanation for alcoholism. AA gives drinkers something to do when they are not drinking. It offers occasions for the soul-satisfying experience of helping someone else. It provides companions who do not drink. And it provides hope for those who need it desperately—the alcoholic and his family—and instant help for the man who wants to get back on the wagon and can't quite make it.

This sounds like a wonderful package of services, and AA is often credited with helping more alcoholics than all the other alcoholism treatments combined. There is no way of knowing whether this is true, since the kind of careful studies needed to show it have not been done. However, most professionals working with alcoholics agree that certainly nothing is lost by encouraging them to attend AA meetings and often much can be gained.

BEHAVIOR THERAPY

There are two kinds of behavior therapy with two different ancestries. One comes from Pavlov, who conditioned dogs. The other comes from B. F. Skinner, who conditioned pigeons. Behavior therapy, in other words, is conditioning therapy by another name.

Pavlov found that if dogs repeatedly heard a bell before eating, eventually bells alone would make them salivate. And if you shocked the dog's foot every time he heard a bell, he would soon respond to bells in the same way he did to shock—by withdrawing the foot.

For many years attempts have been made to condition alcoholics to dislike alcohol. Alcoholics are asked to taste or smell alcohol just before a preadministered drug makes them nauseated. Repeated pairing of alcohol and nausea results in a conditioned response—after a while alcohol alone makes them nauseated. Thereafter, it is hoped, the smell or taste of alcohol will cause nausea and discourage drinking.

Instead of pairing alcohol with nausea, other therapists have associated it with pain, shocking patients just after they drink, or they have associated it with the panic experienced from not being able to breathe by giving them a drug that causes very brief respiratory paralysis. Others have trained patients to imagine unpleasant effects from drinking, hoping to set up a conditioned response without causing so much actual distress.

Does it work? Some degree of conditioning is usually established, but it is uncertain how long the conditioning lasts. The largest study that involved conditioning alcoholics was conducted some forty-five years ago in Seattle, Washington. More than 4,000 patients conditioned to feel nauseated when exposed to alcohol were studied ten to fifteen years after treatment. Half were abstinent, which is an impressive recovery rate compared to other treatments. The patients who did best had booster sessions—that is, they came back to the clinic after the initial treatment to repeat the conditioning procedure. Of those who had booster sessions, 90 percent were abstinent. Based on this study, the nausea treatment for alcoholism would seem an outstanding success. Why hasn't it been universally accepted?

One reason is that the results can be attributed to factors other than the conditioning. The patients in the study were a special group. Generally they were well educated, had jobs, and were well off financially. They may not have received the treatment otherwise, since the clinic where they were treated is private and costs money. Indeed, the patients who did best, it turned out, had the most money. Studies of alcoholics have often shown that certain subject characteristics are more predictive of successful-treatment outcome than the type of treatment administered. These include job stability, living with a relative, absence of a criminal record, and living in a rural community. In the Seattle study there was no control group that did not receive conditioning therapy. It is possible that this select group of patients, many having characteristics that favor a good outcome, would have done as well without conditioning.

Furthermore, in conditioning treatments, motivation is important. Treatment is voluntary and involves acute physical discomfort, so presumably few would consent to undergo the therapy who were not strongly motivated to stop drinking.

The Seattle study makes this point graphically clear. Those who came back for booster sessions did better than those who didn't, but another group did better still: those who *wanted* to come back but couldn't because they lived too far from the hospital. All of these people remained abstinent.

Other studies of the Pavlovian type of therapy for alcoholism—including chemical, electrical, and verbal conditioning—have been less ambitious and the results have been mixed. To the extent they seemed to help, success may be attributed to factors unrelated to conditioning, such as patient selection, patient cooperation, and so on. Another factor also may promote at least short-term success. It is called the Hawthorne effect and refers to the enthusiasm therapists often have for any treatment that is new. This enthusiasm may be infectious, and patients who are enthusiastic themselves about a particular treatment may do somewhat better, for a time, than those who are neutral or unenthusiastic.

The same considerations apply to therapy based on the work of B. F. Skinner, called operant conditioning. There is a large scientific literature based on this work, but the basic ideas are simple. People behave like pigeons in the sense that they do things which are rewarded and avoid doing things which are punished. This has led to a type of treatment known as token economy. Anything a patient does that is believed good for him is rewarded (often literally with tokens which are exchangeable for food, money, and other desirable things). Anything he does that is bad for him is punished (usually by simply withholding the reward). In this manner an attempt is made to "shape" the behavior of patients in directions that are beneficial to them, with the hope that the new behavior—abstinence or controlled drinking—will permanently replace the less beneficial kind.

Does this actually happen? Most treatments of this kind take place in institutional settings, and whether the new behavior brought about in the institution "sticks" in the outside world is not known.

One final word needs to be said about conditioning therapies. By the time they seek professional help, alcoholics have already suffered bitterly from their drinking, but this has not deterred them from continuing to drink. Being made to vomit or having their hand shocked by a friendly therapist is incomparably less excruciating than the physical and mental anguish that alcoholics normally experience: the morning heaves, the shakes, the crushing weight of conscience. There is a sizable delay, to be sure, between the drinking and the anguish, and for conditioning in the literal sense to occur the delay should be shorter. But the effects of heavy drinking are *so* punishing that one would expect some kind of deterrent effect. After all, some people get sick from a pork chop and thereafter avoid pork chops. The need to drink must be compelling indeed, given the infinitely greater misery that comes from drinking. This is what addiction really means, as was movingly described by William James (who incidentally had a brother who was alcoholic):

> The craving for drink in real dipsomaniacs [drunkards] is of a strength of which normal persons can form no conception. "Were a keg of rum in one corner of a room and were a cannon constantly discharging balls between me and it, I could not refrain from passing before that cannon in order to get the rum." "If a bottle of brandy stood at one hand and the pit of hell yawned at the other, and I were convinced that I should be pushed in as sure as I took one glass, I could not refrain." Such statements abound in dipsomaniacs' mouths.

James then gives two case histories:

> A few years ago a tippler was put into an almshouse. Within a few days he had devised various expedients to procure rum, but failed. At length, however, he hit upon one which was successful. He went into the wood-yard of the establishment, placed one hand upon the block, and with an axe in the other, struck it off at a single blow. With the stump raised and streaming he ran into the house and cried, "Get some rum! my hand is off!" In the confusion and bustle of the occasion a bowl of rum was brought, into which he plunged the bleeding member

> of his body, then raising the bowl to his mouth, drank freely, and exultingly exclaimed, "Now I am satisfied." [There also was the] man who, while under treatment for inebriety, during four weeks secretly drank the alcohol from six jars containing morbid specimens. On asking him why he had committed this loathsome act, he replied: "Sir, it is as impossible for me to control this diseased appetite as it is for me to control the pulsations of my heart."

To control "this diseased appetite" with a few sessions of conditioning therapy seems a little like attacking an elephant with a pea shooter. However, one thing can be said in favor of conditioning therapy: It is inexpensive, probably does no harm, and arises from a scientific tradition that emphasizes evidence more than faith.

DRUG THERAPY

In Sears Roebuck catalogs at the turn of the century two pages were devoted to drug therapies for morphine addiction and alcoholism, respectively. The drug being sold for morphine addiction consisted mainly of alcohol; a good part of the drug for alcoholism consisted of tincture of opium, a relative of morphine. Whether morphine addicts became alcoholics as a result of the treatment, or vice versa, is not known, but it illustrates the long history of giving one drug that affects mood and behavior to relieve the effects of another drug that affects mood and behavior. Substitution therapy has reached its pinnacle, perhaps, with the recent widespread, officially sanctioned, and possibly useful substitution of methadone (an addicting substance like heroin) for heroin. Heroin itself was introduced at the turn of the century as a "heroic" cure for morphine addiction and was also believed to be useful for alcoholism. It has not done much for either condition.

Drugs are still widely prescribed to alcoholics. They mainly consist of drugs for anxiety, such as Xanax and Ativan, and drugs for depression. There are no studies indicating they offer long-term help for alcoholics. The antianxiety drugs have some effects similar to those of alcohol—they calm and relax—and are useful in relieving the jitteriness that follows heavy drinking, so that they may be useful in stopping a drinking bout. Whether they stop the *resumption* of drinking—the test of a drug's true worth in treating alcoholism—is debatable, and many clinicians feel they do not.

These drugs are sometimes used in excess by alcoholics, and sometimes in combination with alcohol. This may not be as harmful as it sounds, since they have a low range of toxicity and few people become addicted in the literal sense of needing increasingly larger amounts and having serious withdrawal symptoms when they stop taking them. Nevertheless, they have obviously contributed little to the management of alcoholism, and some clinicians feel strongly that they should not be given to alcoholics for extended periods.

Nor is there evidence that antidepressant medications are useful in the treatment of alcoholism, although some alcoholics do become seriously depressed and antidepressant drugs may then be indicated for their depression.

In the early 1970s, lithium, a drug useful in the treatment of mania, was given to alcoholics, and early reports indicated that some benefited from it. One clinical observation that lends support to the possibility that lithium might help alcoholics is that people with manic-depressive disease often drink more when they are manic than when they are depressed. However, since early reports indicating that a particular treatment is useful are often refuted by later reports, lithium therapy was viewed with both interest and skepticism.

In 1987, Jan Fawcett and colleagues found that more alcoholics taking lithium remained abstinent for a year than those taking a placebo, confirming the earlier studies. Methodological problems involving patient cooperation made the findings less than conclusive, but should encourage further study. Whether the patients were manic or depressed was not a factor in the outcome.

From a theoretical viewpoint, it is interesting that antianxiety and antidepressant drugs do not seem to deter alcoholics from using alcohol. There is ample evidence that these drugs do indeed relieve anxiety and depression, and if alcoholics drink because they feel anxious and depressed, one would assume that the drugs would substitute for alcohol more than they seem to do. This brings up the old question considered earlier in this book, namely, are addictions specific? Are people who are vulnerable to alcohol abuse only vulnerable to alcohol abuse? As said before, there is evidence both supporting and opposing this possibility.

Perhaps the drug most commonly prescribed for alcoholism over the past twenty-five years is one that has no effect on anxiety or depression or apparently anything else unless combined with alcohol. This, of course, is Antabuse.

The drug makes people physically ill when they drink. When it was first used in the early 1950s, Antabuse got a bad name for two reasons. First, like most highly touted treatments, it was not the panacea its enthusiastic supporters had hoped it would be. Second, some people taking the drug died after drinking. It was later learned that the drug could be given in smaller amounts and still produce an unpleasant reaction when combined with alcohol, but death was exceedingly rare.

Partly because of the bad reputation it obtained in the early years, it has perhaps been underused since then. The more dogmatic members of AA view Antabuse as somehow incompatible with the spirit of AA, and many alcoholics resist taking the drug on the grounds that it is a "crutch."

Antabuse has not been entirely popular with doctors for another reason. Many still believe they must give an Antabuse "challenge" test before prescribing the drug for indefinite periods. This test consists of giving the patient Antabuse for a few days and then giving him a small amount of alcohol to demonstrate what an Antabuse reaction is like. The Antabuse challenge test is no longer considered necessary or even desirable. Patients can be *told* what the effects of Antabuse will be and this will have the same effect. One awkward aspect of the challenge test is that some patients have no reaction when given the alcohol, simply because people react very differently to both alcohol and Antabuse and the cautious doses of alcohol administered are too small to produce an effect.

The main problem with Antabuse, however, is not that patients drink after taking the drug but that they stop taking the drug because they "forget" to take it or convince themselves the drug is causing side effects, such as impotency. A way to obviate this problem, at least temporarily, is described below.

SOMETHING THAT WORKS, PROVIDED...

There is an approach to treating alcoholism that works every time, given one stipulation: The patient must do what the doctor says. In this case he must do only one thing: come to the office every three or four days.

Doctors cannot help patients, as a rule, who refuse to do what they say, so there is nothing unusual about the stipulation. Why every three or four days? Because the effects of Antabuse last up to five days after a person takes it. If the patient takes Antabuse in the office, in the presence of the doctor, they both know he will not drink for up to five days. They have bought time, a precious thing in the treatment of alcoholism.

This approach involves other things besides Antabuse, but Antabuse makes the other things possible. First it gives hope, and hope by the time the alcoholic sees a doctor is often in short supply. He feels his case is hopeless, his family feels it is hopeless, and often the doctor feels it is hopeless. With this approach the doctor can say, "I can help you with your drinking problem" and mean it. He doesn't mean he can help him forever (forever is a long time) and it doesn't mean the patient won't still be unhappy or that he will become a new man. It merely means he will not drink as long as he comes to the office every three or four days and takes the Antabuse. Properly warned, he won't drink unless he is crazy or stupid, and if either is the case, he probably should not be given Antabuse.

On the first visit the doctor can say something like this:

"Your problem, or at least your immediate problem, is that you have trouble controlling your drinking. Let *me* take charge; let me control your drinking for a time. This will be my responsibility. Come in, take the pill, and then we can deal with other things.

"I want you to stop drinking for a month. [At this point the doctor makes a note in his desk calendar to remind himself when the patient will have taken Antabuse for a month.] After that we can discuss whether you want to continue taking the pill. It will be your decision.

"You need to stop for a month for two reasons. First, I need to know whether there is anything wrong with you besides drinking too much. You may have another problem that I can treat, such as a depression, but I won't be able to find out until you stop drinking for at least several weeks. Alcohol itself makes people depressed and anxious, and mimics all kinds of psychiatric illnesses.

"Second, I want you to stop drinking for a month to have a chance to see that life is bearable—sometimes just barely bearable—without alcohol. Millions of people don't drink and manage. You can manage too, but you haven't had a chance recently to discover this."

F. Scott Fitzgerald complained that he could never get sober long enough to tolerate sobriety, and at least this much can be achieved with the present approach.

It is important for the patient to see the doctor—or whatever professional is responsible for his care—whenever he comes for the pill. Patients as a rule want to please their doctors; this is probably why they are more punctual in keeping office appointments than doctors are in seeing them. In the beginning the patient may be coming, in part, as a kind of favor to the doctor.

The visits can be as brief or as long as time permits. The essential thing is that rapport be established, that the patient believe something is being done to help him, and that he stay on the wagon (he has no choice if he lives up to his part of the doctor-patient contract). Brief, frequent visits can accomplish these things.

The emphasis during the visits should be not on the pill but on problems most alcoholics face when they stop drinking. The major problem is finding out what to do with all the time that has suddenly become available now that drinking can no longer fill it. Boredom is the curse of the nondrinking drinking man. For years, most of the pleasurable things in his life have been associated with drinking: food, sex, companionship, fishing, Sunday afternoon football. Without alcohol these things lose some of their attraction. Who can enjoy French cooking without wine, tacos without beer, or business luncheons without martinis? The alcoholic is sure he cannot. He tends to withdraw, brood, feel sorry for himself.

The therapist may help him find substitute pleasures—hobbies, social activities not revolving around alcohol, anything that kills time and may give some satisfaction, if not anything as satisfying as a boozy glow. In time he may find these things for himself, but meanwhile life can be awfully monotonous.

Also the patient can bring up problems of living that tend to accumulate when a person has drunk a lot. People usually feel better when they talk about problems, particularly when the listener is warm and friendly and doesn't butt into the conversation by talking about his own problems. The therapist can help by listening even if he cannot solve the problems.

If he is a psychiatrist, he can also do a thorough psychiatric examination, looking for something other than drinking to diagnose and treat. Occasionally—not often—alcoholics turn out to have a depressive illness, anxiety neurosis, or other psychiatric condition.

One thing the therapist can do is help the patient accept his alcoholism. This is sometimes difficult. Alcoholics have spent most of their drinking careers persuading themselves and others that they do not have a drinking problem. The habit of self-deception, set and hardened over so many years, is hard to break. William James describes this habit with his usual verve and concludes that the alcoholic's salvation begins with breaking it:

> How many excuses does the drunkard find when each new temptation comes! Others are drinking and it would be churlishness to refuse; or it is but to enable him to sleep, or just to get through this job of work; or it isn't drinking, it is because he feels so cold; or it is Christmas-day; or it is a means of stimulating him to make a more powerful resolution in favor of abstinence than any he has hitherto made; or it is just this once, and once doesn't count . . . it is, in fact, anything you like except *being a drunkard.* But if . . . through thick and thin he holds to it that he is a drunkard and nothing else, he is not likely to remain one long. The effort by which he succeeds in keeping the right *name* unwaveringly present to his mind proves to be his saving moral act.

After a month of taking the pill and talking about problems, what happens then? The patient and doctor renegotiate. Almost invariably, in my experience, the patient decides to take the pill for another month. The doctor says okay, and this is the first step in a process that must occur if the patient is going to recover: acceptance of personal responsibility for control of his drinking.

Proceeding on a month-to-month basis is a variation on the AA principle that an alcoholic should take each day as it comes. For years, alcohol has been the most important thing in the alcoholic's life, or close to it. To be told he can never drink again is about as depressing as anything he can hear. It may not even be true. Studies indicate that a small percentage of alcoholics return to "normal" drinking for long periods. "Controlled" drinking is probably a better term than "normal" drinking, since alcoholics continue to invest alcohol with a significance that would never occur to the truly normal drinker.

Many people, especially among AA members, reject the notion that alcoholics can ever drink normally. If alcoholism is defined as a permanent inability to drink normally, then obviously any person able to drink normally for a long period was never an alcoholic in the first place. The issue is really a definitional one, and those few alcoholics who reported sustained periods of controlled drinking in the studies were at any rate considered alcoholic when they *weren't* drinking normally. Most clinicians would agree that it is a mistake to encourage an alcoholic to believe he can ever again drink normally, but on the other hand telling him he can never drink again seems unnecessary and may not be true in every case.

When does treatment end? The minimum period is one month because that is the basis for the doctor–patient contract agreed upon in advance. Ideally, however, the treatment should continue for at least six months, with the patient himself making the decision to continue taking Antabuse on a month-to-month basis. Why six months? Because there is evidence that most alcoholics who begin drinking again do so within the first six months following abstention.

A general rule applies here: The longer a patient goes without drinking at all, the shorter the relapse if a relapse occurs. It takes time to adapt to a sober way of life. Both the doctor and patient should be prepared for relapses. Alcoholism, by definition, is a chronic relapsing condition, although relapses are not inevitable. It resembles manic-depressive disease in this regard and also has similarities to such chronic medical illnesses as diabetes and multiple sclerosis. When the alcoholic has a relapse, his physician often feels resentful. When his diabetic patient has a relapse because he failed to take insulin, the doctor tends to be more understanding. The reason for this inconsistency is not clear.

Three objections have been raised concerning the above approach to treating alcoholism. The treatment is said to be based on fear, namely, the fear of getting sick, and fear is held to be one of the least desirable forms of motivation. This is debatable. Fear may be the *only* reason some alcoholics stop drinking. There is evidence that internists have somewhat better success in treating alcoholics than psychiatrists do, and the reason may be that they are in a better position to frighten the patient. They have merely to examine his liver and tell him he may be dead in a year if he keeps on drinking. Innumerable alcoholics have stopped drinking because they were told something like this. Others have stopped because they were afraid of losing their wives or jobs. It is probably no coincidence that the hardest alcoholics to treat are those who have little to lose, those who have already lost their wives, jobs, and health. They have no hope of regaining these. All they have left to lose is their life, and by now living has little appeal. Probably the most effective alcoholism treatment programs are run by industries, where the patient is an employee and his job depends on staying sober. The quotation at the beginning of this chapter is from Gorham's fictional study of an alcoholic, *Carlotta McBride.* As Gorham says, alcohol works. It has worked for the alcoholic for many years. Unless he is very much afraid of *something,* he probably will not give it up.

The second objection to the approach outlined here is that the patient becomes too dependent on a personal relationship with an authority figure, the physician, which must end at some point. In the treatment of alcoholism, the goal is not so much a lifetime cure—although this often happens—as it is to bring about improvement. If the patient stays sober for longer periods after treatment than he did before, the treatment has been at least a limited success. The physician in any case should discourage a dependent relationship. He can insist upon the patient's taking the pill and staying dry for a month (realizing that a month is an arbitrary unit of time and any fixed interval will do), but after that the patient has to realize that he himself has the ultimate responsibility for the control of his drinking.

Finally, the complaint is heard that this approach does not get at the root of the problem; it does not explain how the patient became an alcoholic. This is true but, in my opinion, no one can explain how a person becomes alcoholic because no one knows the cause of alcoholism. Doctors sometimes blame the patient's upbringing and patients often blame everyday stresses. There is no way to validate either explanation. There is probably no harm in telling the patient that his condition remains a medical mystery. And despite everything said in this book, it is still premature to say that he *inherited* his disease.

However, if it is ever shown conclusively that some forms of alcoholism are influenced by heredity, this would not make the prognosis less favorable or the treatment less helpful. Sometimes, when evidence for a genetic factor is presented, you hear the following: "But if it is genetic, then you can't do anything about it." It should be noted that diabetes is almost certainly a genetic disorder and there are excellent treatments for diabetes.

The most serious objection to the above approach in treating alcoholism is that it has not been studied and there is no evidence that it is better than any other treatment. For this, the writer apologizes and promises he will do whatever he can to correct the situation.

SUGGESTIONS FOR PHYSICIANS

Before alcoholism can be treated, it first must be recognized. Physicians are in a particularly good position to identify a drinking problem early: They can do a physical examination and order laboratory tests. Here are some hints of a drinking problem:

1. Arcus senilis—a ring-like opacity of the cornea—occurs commonly with age, causes no visual disturbance, and is considered an innocent condition. The ring forms from fatty material in the blood. Alcohol increases fat in the blood, and more alcoholics are reported to have the ring than others their age.
2. A red nose (acne rosacea) suggests the owner has a weakness for alcoholic beverages. Often, however, people with red noses are teetotalers, or even rabid prohibitionists, and resent the insinuation.
3. Red palms (palmar erythema) are also suggestive, but not diagnostic, of alcoholism.

4. Cigarette burns between the index and middle fingers or on the chest, and contusions and bruises, should raise suspicions of alcoholic stupor.
5. Painless enlargement of the liver may suggest a larger alcohol intake than the liver can cope with. Severe, constant upper abdominal pain and tenderness radiating to the back indicates pancreatic inflammation, and alcohol sometimes is the cause.
6. Reduced sensation and weakness in the feet and legs may occur from excessive drinking.
7. Laboratory tests provide other clues. After heavy drinking, more than half of alcoholics have increased amounts of gamma-glutamyl transpeptidase (GGT) in their blood, which is unusual in nonalcoholics. After GGT, elevations in the following blood tests are most often associated with heavy drinking: mean corpuscular volume, uric acid, triglycerides, aspartate aminotransferase, urea, and acetate.

If suspicious, the physician can administer the CAGE test. It consists of four questions:

Have you ever:

- felt the need to *C*ut down on drinking?
- felt *A*nnoyed by criticism of your drinking?
- had *G*uilty feelings about drinking?
- taken a morning *E*ye-opener?

Patients who say "yes" to all four are usually alcoholic by any definition, according to one study.

WHAT CAN THE FAMILY DO?

First, it can recognize the problem when there is one. There often is great reluctance to do so. Wives sometimes get maternal gratification from caring for drinking husbands—it makes them feel superior, like the head of the family. The children may prefer a tipsy, happy daddy to a sour, sober father. The problem has to be seen as a problem—but then what?

Nothing is more frustrating for someone in the helping business—doctor, social worker—than to get a call from the spouse: "John is drinking too much. What can I do?"

"Well, have him come see me."

"But he won't. He doesn't think he has a drinking problem. I'm desperate. What should I do?"

Call the police? What can the police do? Usually nothing. Drinking is not a crime. Wife-beating is, and that is when the police may intervene, but they usually cannot help with the drinking.

Nag? Nagging just provides another reason to drink.

Threaten? Well, yes, sometimes. If the last straw is really the last straw, it may be a good idea to say so. Sometimes people do stop drinking because a husband or wife threatens to leave them. Coercion sometimes works. But what if it isn't the last straw? If threats or importunings don't work, what will? Family life poses few problems as painful as this one. How the spouse handles it says a good deal about the spouse, and also about the role of the sexes in dealing with each other's problems.

There are infinite variations, of course. "Happy families are all alike," wrote Tolstoy. "Every unhappy family is unhappy in its own way." In their own ways, husbands and wives play out the roles society and their own unique personalities assign to them, and usually there is not as much "choice" as people like to believe. Advice, even from the wisest adviser, may be bad advice simply because family relations are tremendously complicated, played out intuitively by and large, and outsiders never know how it *really* is.

Unequivocal advice to be tender or tough—to leave him or not—is usually best not given, and most people don't listen anyway.

Here are some general principles to remember:

1. The alcoholic must face the consequences of his behavior.

The family often tries to protect him from these consequences and shouldn't.

Don't pick up the pieces. If he passes out, leave him there. If he throws up, let him clean it up the next morning. If he doesn't remember how the window was broken, tell him later. Be matter of fact. Don't pretend it was funny. Don't say I told you so. (Blackouts are scary. Sometimes people stop drinking because of them. Don't let him forget he forgot.)

Don't buy him a drink.

Don't call the boss to say he has the flu (a hard rule to follow when the family depends on the income).

Don't bail him out of jail—or anything else. Let *him* explain—not you. Let *him* apologize—not you.

Stop trying to control his drinking behavior. You can't anyway. You are as powerless in this regard as he is. Stop playing games. Stop hiding bottles. Stop pouring booze down drains. Stop organizing the family routine around his drinking; shortening the cocktail hour won't help. Stop babying him. Allow *him* to be responsible for *his* behavior. Love the sinner but not the sin.

2. Don't preach. It doesn't help.
3. Keep up hope. Many alcoholics just up and recover, with help or without.
4. Save yourself.

Finally, try the telephone book. AA usually has a listing, and so does Al-Anon. AA usually won't send someone to the house—the alcoholic must first ask—but the spouse can go to Al-Anon and the kids can go to Alateen. If there is a National Council on Alcoholism chapter in town, it can direct you to these groups, as can a minister. Physicians often are less helpful; studies repeatedly show that physicians are woefully ignorant of the subject.

The alcoholic is isolated and ashamed. The alcoholic's family is isolated and ashamed. The kids don't invite friends over because father may be drunk. Bowling and Saturday night bridge become things of the past.

Isolated, ashamed, and bewildered, the family thinks this problem has never happened to anyone else. It has. Al-Anon and Alateen are opportunities to find this out. They are opportunities to learn how others deal with it. They are opportunities to learn how you may be contributing to the problem yourself without knowing it.

Should the family directly confront the alcoholic—lay it down straight and simple? Confrontation has become a big word in the treatment field. Some favor it more than others. It seemed to work with Betty Ford. Jerry and the children descended on her one day and said she was drinking too much. They brought along the head of a nearby alcoholism hospital. He gave her a copy of the book *Alcoholics Anonymous* and suggested she substitute the words "chemical dependence" in the book for "alcoholism." "I was in shock," writes Mrs. Ford in her autobiography. She cried. She was enraged. But she went in the hospital, stayed a month, and returned home free of her chemical dependence. Forever? In Mrs. Ford's case, it seems quite likely.

Confrontation is tricky. It works in some families, fails in others. Al-Anon, Alateen, or a good alcoholism counselor can help a family decide whether to try.

CHAPTER NINE

The Key Questions

This chapter ends our fairly wide-ranging discussion of alcohol and alcoholism with an attempt at drawing some conclusions. Basically, we've addressed four questions. Here are the key questions and what we can say, briefly, about each of them.

IS ALCOHOLISM HEREDITARY?

For centuries physicians and others have believed that alcoholism runs in families: that the cause was heredity, although the drinking itself was a *vice.* Only recently have serious scientific efforts been made to determine just what role heredity does play.

The basic problem in separating heredity from environmental influences (the other prime suspect) is that most people are raised by their biological relatives, usually their parents. The influences of homes and chromosomes on the final product—the person—quickly become virtually inseparable.

Some authorities believe that the contributions of heredity and environment can never be fully disentangled. Writing about the origins of intelligence, Donald Hebb warned of the dangers of regarding either heredity or environment as the primary cause: "Each is *fully* necessary . . . to ask how much heredity contributes to intelligence is like asking how much the width of a field contributes to its area."

Therefore, our question becomes:

CAN WE PROVE THAT ALCOHOLISM IS HEREDITARY?

Despite Hebb's caution, other authorities believe that the solution to the nature–nurture problem is worth pursuing. If their work indicates that a biological or genetic factor is important, scientists will have a clear direction for their research on cures or treatments.

Today, researchers are engaged in a widespread search for the "genetic markers" that could establish an inherited basis for alcoholism. Their search takes several forms. One can look for variations in the genetic material (DNA) that can be associated with alcoholism. A chromosomal marker for alcoholism may be—if found—the definitive evidence for the role of heredity that people are looking for.

The technology for this type of research has just been developed, and the search just begun. At this point, finding a chromosomal marker is a little like looking for a needle in a haystack; but the "needle" has been found, apparently, for some other illnesses (Huntington's chorea, for example, and, possibly, for depression).

Researchers are also looking at other types of markers—like characteristic differences between the children of alcoholics and the children of nonalcoholics—that may *predict* alcoholism.

For example, if some of the children of alcoholics show brain-wave abnormalities that seem to predict future alcoholism, and this correlation proves to be strong and consistent, it would provide good evidence for a constitutional, if not genetic, contributing cause for the illness.

(*Genetic* means that genes made the difference; *constitutional* means that you were born that way. These are not always the same thing—more than genes go into the way you are born; for example, you can be affected by certain events that occurred during your mother's pregnancy.)

At present, while there is *evidence* for a hereditary factor in *some* cases of alcoholism, the evidence is not strong enough for definite conclusions. It's important to remember the cautionary tale of kuru (from chapter 6): Everyone thought that kuru was inherited; but it turned out to have a dietary basis. We should also pay close attention to alternative explanations for what appears to be the effect of heredity.

For example, if significant numbers of adoptees become alcoholic even when they've been separated from their alcoholic biological parents, it could mean that alcoholism is inherited, but it could also mean that the cause is prenatal: Perhaps the mother drank during pregnancy, and this increased the likelihood of her offspring developing alcoholism. There is no evidence that this particular chain of events occurs, and most authorities dismiss it as improbable, but alternatives are the very stuff of science, and they should never be rejected until conclusively disproved.

At the moment, even though the answer to our question is, *"No, we can't prove it,"* the evidence keeps mounting that heredity plays an extremely important role in the development of alcoholism.

WHAT *DO* WE KNOW?

Alcoholism certainly runs in families. As we've seen, between 20–25 percent of the sons of alcoholics (and about 5 percent of the daughters) become alcoholic themselves. These rates are about four to five times greater than for the general population.

Alcoholism runs in families even when the children are separated from the alcoholic biological parents and raised by nonalcoholic adoptive parents. Investigators in three countries have supported this with remarkably similar findings.

There can be only one conclusion: Some people begin life with a susceptibility—with the emphasis on *susceptibility*—to alcoholism.

This seems to establish that alcoholism—in some people—does not require exposure to alcoholic behavior in others. It would therefore appear not to be "learned" behavior, but to be dependent on other factors, which may be genetic (involving variations in DNA) or constitutional.

Some people should also be emphasized. Not all alcoholics have a family history of alcoholism. There may be two kinds of alcoholism: a familial type, possibly genetic in origin, and a nonfamilial type.

"Familial" alcoholics appear to become alcoholic at an earlier age and develop a more severe form of the illness. They may also be harder to treat.

"Nonfamilial" alcoholics may be alcoholic in a different way. They may be drinking to treat some other problem they have: anxiety, depression, a familial tremor. Perhaps, as often happens, the self-treatment got out of hand.

Maybe these alcoholics drink to feel better—and it works, for a while. Alcoholics experience a rapid cyclic effect from alcohol characterized by brief periods of euphoria followed by unpleasant feelings—which can be relieved by more alcohol. They begin to drink more than ever: the familiar vicious cycle characteristic of all addictive drugs, including caffeine and nicotine.

As Hebb's warning predicted, it has proven to be extremely difficult to separate the influences of heredity and environment. Studies of the interaction between environment and constitutional or genetic factors in alcoholism are in their infancy—and it's still too early to draw definite conclusions.

However, further analysis of some of the adoption studies (our best evidence for the influence of heredity) suggests that environmental factors may determine whether a biological tendency toward alcoholism becomes manifest.

No one denies that environment is an important factor. To begin with, the disease can't develop unless alcohol is available. There are typically very low rates of alcoholism in countries where alcohol is relatively unavailable for religious reasons (e.g., Islamic countries).

Also, parental attitudes toward drinking surely influence the drinking behavior of their children, although it's difficult to say what attitudes produce what behaviors.

Much has been written about the problems that the children of alcoholics experience because of their parents' drinking. However, there is little evidence that the children of alcoholics suffer any more that the children of people with other psychiatric illnesses—and either environmental, constitutional, *or* genetic factors may explain their problems. This issue remains unresolved.

The wide differences in drinking patterns between different Judeo-Christian cultures are further evidence for an environmental explanation for alcoholism. Irish immigrants have a notoriously high rate of alcoholism, while Jews have a low rate. Genetic factors don't seem to explain these differences.

Millions of people seem to be more or less protected from abusive drinking: They become ill after taking small amounts of alcohol. Apparently many Orientals have this alcohol-related "problem," as do many Caucasian women. This may help explain the low rate of alcoholism among Orientals, and the fact that Caucasian women are less likely to develop alcoholism than Caucasian males. In both cases a physiological aversion to alcohol may be as powerful a deterrent to heavy drinking as cultural sanctions, but it is unknown where the influence of one starts and the other leaves off.

To whatever extent heredity does influence alcoholism, the chemical mechanism by which it works is unknown. There are some interesting leads:

- Researchers have discovered that alcohol produces tiny amounts of a morphine-like compound in the brain; alcoholics may be genetically predisposed to produce larger than usual quantities of these (possibly addictive) substances.
- The levels of certain chemical mediators in the brain, such as serotonin, are temporarily increased and then decreased by alcohol, a cyclic effect that roughly corresponds to the drunk-hangover cycle.
- Finally, there have been reports that alcoholism can cause changes in the receptors in the brain that bind to tranquilizers and opiates.

All of this research is at a very early stage and reflects the high level of interest in finding biological—and possibly genetic—explanations for alcoholism.

WHY IS THIS INFORMATION IMPORTANT?

In medicine, finding the cause of a condition is usually an essential step in finding a cure. Once the tubercle bacillis was found, a cure for tuberculosis followed—and with the AIDS virus apparently identified, a vaccine for AIDS is a distinct possibility.

Knowing how alcoholism is transmitted from generation to generation, either at a biological or cultural level, will greatly increase the chance that we'll develop effective prevention or a cure.

For example, the rapid drunk-hangover cycle discussed above may have a biological basis. If that basis can be discovered, it may be possible to dampen the cycle by chemical means—reducing the alcoholic's need to drink.

This may sound utopian, but fifty years ago antibiotics sounded utopian. In any case, recent evidence that alcoholism may be inherited offers hope for future victims. Illnesses influenced by heredity, in general, are just as treatable as those that are not. One simply needs to know something about how heredity does its work.

From a more immediate standpoint, the children of alcoholics should be aware that they stand an increased risk of becoming alcoholic. A recent poll showed that only 5 percent of the respondents—including children of alcoholics—realized that alcoholism ran in families.

Heredity, somebody said, is what you believe in if your kid makes straight A's. Based on the present evidence, it's also what you should believe in if members of your family have had trouble with alcoholism. Many children of alcoholics who are aware of their vulnerability have started—for this very good reason—taking special precautions with alcohol.

Education is often criticized as being a frail need for changing pleasurable or recreational behavior. However, the percentage of smokers in the population has decreased from 50 percent to 25 percent over the past two decades, and education deserves at least part of the credit. Educating the families of alcoholics about the risks they run may be the most powerful tool for prevention that we have today.

Notes and References

CHAPTER ONE

p. 9

An invaluable source in reviewing the physical properties of alcohol and alcohol metabolism is a multivolume work entitled *The Biology of Alcoholism,* edited by Benjamin Kissin and Henri Begleiter and published by Plenum Press. Its 103 chapters are spread over 4,436 pages; it weighs eighteen lbs, is ten inches tall and fourteen inches in length. For anything you might want to know about alcohol and alcoholism but were afraid to ask, this is the place to go. The title is misleading, since behavioral and social aspects of alcoholism are dealt with as much as biology. The seventh volume appeared in 1985, but two more are in the works.

Another encyclopedic approach to the subject consists of five volumes (as of 1987) called *Recent Developments in Alcoholism,* edited by Mark Galanter and published by Plenum Press. Both sets are indispensable for serious students of the subject.

For those daunted by multivolume sets (or too poor to afford them), I warmly recommend two volumes for general background. One is the *Encyclopedic Handbook of Alcoholism,* edited by Pattison and Kaufman (Gardner Press, 1982). The other is *Alcohol: Use and Abuse in America* by Mendelson and Mello (Little, Brown, 1985). The latter is particularly helpful in providing a wise perspective on the American love–hate affair with alcohol dating back to the Pilgrims.

p. 9
To say that ethyl alcohol is "just water with an ethyl group" is like saying that glucose is "just" water latched onto half a dozen carbon atoms. A few carbon atoms can make a lot of difference. However, alcohol does behave similarly to water, travels everywhere that water travels, and, because of its water-like properties, can be accommodated by the body in vastly greater amounts than any other drug. A person's blood can consist of half a percent of alcohol without producing death or even unconsciousness.

p. 9
The concentration of congeners varies from beverage to beverage, from brand to brand, and even from bottle to bottle of the same brand. For example, not all brands (or bottles) of vodka contain relatively high amounts of methanol, but some do. Russian vodka, the favorite of many vodka connoisseurs, sometimes contains considerable methanol, although, again, probably not enough to be harmful.

p. 10
The alcohol content of most distilled alcoholic beverages is expressed in degrees of proof. This term probably developed from the seventeenth-century English custom of estimating alcohol content by moistening gunpowder with the beverage and applying a flame to the mixture. The lowest alcohol concentration that would allow ignition—a concentration of about 57 percent alcohol by volume—was considered to be "proof spirits." British and Canadian regulations are still based on this yardstick; a concentration of 57.35 percent alcohol is considered to be "proof spirits," while other concentrations are described as "over" or "under" proof.

p. 10
The reason that fortified wines and distilled spirits do not "sour" in the open air is that they have concentrations of alcohol above 12 or 13 percent, which is as lethal for bacteria as it is for yeast.

p. 10
A minute amount of ethyl alcohol is produced in the gastrointestinal tract by bacteria, and perhaps this accounts for alcohol dehydrogenase in the liver. Infinitesimal amounts of alcohol may also be produced by normal metabolic processes in the body. If these sources are the reason that the alcohol enzyme is present in such large quantities, it is clearly a case of biological overkill.

p. 12
Alcohol interferes with the absorption of fats, vitamins, and other important dietary constituents. Because of this, it has been suggested that perhaps the baboons were not well nourished despite having an adequate nutritional intake. Therefore, even this important study has not really resolved the issue of whether alcohol, nutrition, or both are responsible for alcohol-related liver disease. The study was written by two leading authorities in the field, Drs. Charles Lieber and Emanuel Rubin, and published in the *New England Journal of Medicine* 290:128–135.

p. 12
Leonard has a good review of alcohol effects on membranes and neurotransmitters in *Alcohol and Alcoholism* 21:325–338.

p. 13
Two 1983 papers reviewed the effects of alcohol on TIQS in rats and humans, respectively: Clow et al. (*Neuropharmacology* 22:-563–565) and Sojquist et al. (*Drug and Alcohol Dependence* 12:15–23). A 1986 paper on TIQS in alcoholics appears in *Alcohol* 3:371–375.

p. 14
Thomson and McMillen examined the possible relationship of serotonin deficiency to alcoholism in *Alcohol* 4:1–5.

p. 14
A 1985 paper by Borg et al. (*Alcohol* 2:415–418) showed that alcohol causes an increase in serotonin levels in alcoholics, followed by a decrease, supporting the theory that serotonin is low in abstinent alcoholics and activated during abuse.

p. 14
Genazzani et al. explored the hypothesis that a deficiency of endorphins underlies alcohol addiction in the *Journal of Endocrinology* 55:583–586. Evidence supporting the hypothesis was presented by Gianoulakis et al. in the *Canadian Journal of Physiological Pharmacology* 61:967–976.

p. 15
For more about the benzodiazepine-deficiency hypothesis, see Freund's paper on biomedical causes of alcohol abuse in *Alcohol* 1:129–131.

p. 16
One explanation for the popularity of combining alcohol with club soda is that the carbonated water facilitates the introduction of alcohol into the small intestine where absorption takes place almost immediately.

p. 16
There may also be a correlation between the rapidity with which a drug is absorbed into the blood and its potential for abuse. For example, fast-acting barbiturates such as Nembutal have an abuse potential far greater than long-acting barbiturates such as phenobarbital.

p. 19
The first investigator to report flushing in Orientals from drinking alcohol was Peter H. Wolff (*Science* 175:149–450, 1972). Studies by the writer and his colleagues at Kansas University have confirmed the observation that about 80 percent of Orientals develop a cutaneous flush after drinking a small quantity of alcohol. Moreover, they found that the flush can be blocked by predosing the subjects with two antihistamines, Benadryl, 50 mg, and Tagamet, 300 mg, thirty minutes before drinking. A paper by Miller et al. (*Journal of Alcohol Studies,* in press) describes the antihistamine findings. A 1987 paper by Goedde and Aagarwal (*Enzyme* 37:29–44) describes the enzymes involved in the flushing response.

p. 19
Psychological set has been compared to a thermostat on a furnace. If the thermostat is already high, the addition of a small amount of heat from other sources may have decisive effects on behavior (windows are raised, sweater removed). The opposite phenomenon also may occur. If a person is already exhilarated, alcohol may make him somewhat less exhilarated; if he is depressed, alcohol may raise his mood. This phenomenon is called the law of initial value and is one more factor that complicates drug effects.

Set may have a partial genetic basis, according to a 1985 study by Gabrielli and Plomin (*Journal of Nervous and Mental Disease* 173:111–114).

p. 21
For a review of the role of alcohol in homicide, see D. W. Goodwin, "Alcohol in Suicide and Homicide," *Quarterly Journal of Studies on Alcohol* 34:144–156.

p. 21
Dr. Jack H. Mendelson and his colleagues at Harvard University have observed that alcoholic subjects permitted to drink alcohol over long periods become clinically depressed. Of course, drinking on an experimental ward is not quite the same as drinking in real-life circumstances, but even in bars one sees glum drinkers.

In *Alcohol and Aggression* (Croom Helm, 1986), Paul Brain makes a strong case for sociocultural explanations for alcohol-related aggression.

p. 22
State-dependent learning could also be called the Charlie Chaplin effect. In *City Lights,* Chaplin rescued a drunk from drowning. Grateful, the drunk took Charlie home and wined and dined him. The next day, sober, the drunk did not remember the experience and kicked Charlie out of the house. On subsequent occasions, when intoxicated, the drunk remembered Charlie's good deed and treated him well, only to forget it again when sober.

How Chaplin knew about the state-dependent phenomenon is

a mystery, unless it is so common that many people know about it and just don't talk about it. In any case, *City Lights* was made in the early thirties, and the first scientific paper about state-dependent learning was published in the mid-thirties.

A recent paper on the subject by Kent et al. (*Journal of Studies on Alcohol* 47:241–243) reported that young men with a propensity for blackouts when drinking also show marked state-dependent effects in a study that required repeated observations of subjects while drunk and sober.

For the first study reporting state-dependent effects in humans, see Goodwin et al. in *Science* 163:1358–1363.

pp. 22–24

The current literature is rife with studies suggesting that modest consumption of alcohol may be beneficial. Beneficial health effects are reviewed in a 1985 paper by Bauman-Baicker in *Drug and Alcohol Dependence* 15:207–227. A review of the psychological benefits by the same author appears in the same journal (pp. 305–322). Evidence that alcohol use may reduce the risk of coronary artery disease comes from two major studies, one conducted by the Kaiser Permanente Medical Group (Klatsky et al., *American Journal of Cardiology* 58:710–714) and the Framingham study (Anderson et al., *Journal of the American Medical Association* 2176–2180).

The idea that alcohol might be good for you is not new. The seventeenth-century Puritan minister Increase Mather praised alcohol; his son Cotton condemned it. For many years some anthropologists have held that civilization might not have occurred if statesmen had not toasted each other rather liberally and thereby overcome their suspicions and hostilities. According to archaeologist Solomon Katz, prehistoric humans discovered that wild wheat and barley soaked in water and left in the open air made a dark, bubbling brew that made whoever drank it feel good. On top of that, the brew made people robust. Dr. Katz argued that the discovery of a way to produce alcohol produced enormous motivation for continuing to go out and collect these grains and develop more potent brews. "Almost invariably," he wrote, "individuals in societies appear to invest enormous amounts of effort and even risk" in the pursuit of mind-altering foods and beverages. All this, he said, led to the sowing, cultivating, and reaping of crops, namely, agriculture. The paper was published in the June 1987 issue of the *Journal of the Museum of Archaeology/Anthropology.*

Finally, the observation that man the tippler tends to outlive his cousins of the animal kingdom was the subject of a little poem by "Anon.," a favorite of nineteenth-century anthologies:

> The dog at 15 cashes in
> Without the aid of rum and gin,
> The cat in milk and water soaks
> And then in 12 short years it croaks,
> The modest, sober, bone-dry hen
> Lay eggs for nogs, then dies at 10.
> All animals are strictly dry:
> They sinless live and swiftly die;
> But sinful, ginful, rum-soaked men
> Survive for three score years and ten.
> And some of them, a very few,
> Stay pickled till they're 92.

pp. 24–25
The data are from a 1985 survey (*Public Health Reports* 101:593–598).

pp. 24–25
"Normal" can be defined in several ways. It can be defined as drinking no more than society deems safe and prudent. Since societies vary in this regard, the definition is not very helpful. According to another definition, normal drinking is drinking less than is required to produce medical, social, or psychological problems. The problem definition also has problems, as is discussed in chapter 2.

Finally, attempts are made from time to time to separate normal from abnormal drinking in terms of quantity of alcohol consumed. A nineteenth-century British physician named Dr. Anstie proclaimed that normal drinking consisted of drinking no more than three ounces of whiskey or half a bottle of table wine or two pints of beer a day (known for years as "Dr. Anstie's limits"). In 1979 a special committee of the Royal College of

Psychiatrists in Great Britain proposed more liberal limits. The committee announced that four pints of beer daily or "four doubles of spirits" or a bottle of wine "constitute reasonable guidelines for the upper limit of drinking." Upper limits vary from person to person, and many would support the more conservative position of Dr. Anstie.

On the other hand, when Bertrand Russell died in his late nineties, it was reported that he had consumed eight double scotches daily for thirty years. He opposed the Vietnam War and thought nuclear bombs were dangerous. This could be construed as evidence of brain damage from the drinking, but this must remain speculative.

p. 25
The urination phenomenon has been observed by the author and by Dr. Jack H. Mendelson and his group at Harvard. Alcohol blocks the action of a pituitary hormone that reduces urinary output, but the resulting diuresis is only temporary and is related either to large quantities of fluid (as in beer drinking) or to a rapidly rising blood–alcohol level.

p. 25
The dry mouth that comes from a night of heavy drinking probably has other causes besides the stringent effect of alcohol, such as altered distribution of water between cellular and extracellular fluids, but the real reason is not understood. In any case, there is no justification for giving intravenous fluids to alcoholics who have just been drinking, unless there is objective evidence of dehydration (usually from vomiting or diarrhea).

CHAPTER TWO

p. 27
Differences between American and French alcoholics are described in Jellinek's *The Disease Concept of Alcoholism* (College and University Press, 1960), a classic in the field.

p. 28
The American Psychiatric Association first published *The Diagnostic and Statistical Manual of Mental Disorders (DSM–III)* in 1980. A revised version *(DSM–III–R)* was published in 1987. The criteria for alcohol dependence in the two volumes are substantially different, although field tests indicate that both sets of criteria identify essentially the same group of alcoholics. The criteria presented in this book are a slightly modified version of those found in *DSM-I-R*.

pp. 33ff
Even about something as objective as problems, there is still disagreement. How many problems? How serious must the problems be? With David and other classical alcoholics, there is no difficulty—their problems are both abundant and serious. But in milder, less advanced, and less typical cases, there is controversy.

This raises a related question: When does alcoholism begin? Fixing the onset of a chronic condition is difficult. In cancer, when does the first cell become malignant? When does the first coronary artery narrow in heart disease? Cancer and heart diseases usually can be diagnosed only after they are far advanced, and the same is true of alcoholism. With some alcoholics, the alcoholism appears to start with the first drink, but if this happens, often it is usually not apparent except in retrospect.

p. 33
The most ambitious study of drinking practices was published in the early 1970s by Don Cahalan, Ira Cissin, Helen M. Crossley, and Robin Room. The data cited here come from their book *Problem Drinkers* (Jossey-Bass, 1970). A similar coast-to-coast study has not been performed since then, so therefore it is uncertain whether the data are still valid.

However, according to the *Sixth Special Report to the U.S. Congress on Alcohol and Health,* prepared by the National Institute on Alcohol Abuse and Alcoholism in 1987, per capita alcohol consumption in the United States has declined significantly since 1980. Consumption during that period dropped from 2.76 gallons of pure alcohol per person over fourteen years of age in 1978 to 2.65 gallons in 1984.

This decline was reflected in a decrease of driving fatalities: Between 1980 and 1984 the proportion of fatally injured drivers who were legally intoxicated dropped from 50 to 43 percent. Mortality for liver cirrhosis declined to the lowest level since 1959.

The number of individuals receiving treatment for alcoholism, however, continued to grow, with a half million reported in treatment in the mid-1980s. This may simply mean there are more treatment facilities, not more alcoholics.

pp. 36–37
Although many alcoholics appear depressed upon entry into an alcoholism treatment program, the depression tends to disappear in most cases after a few weeks of sobriety. Overall et al. documents this in *Alcoholism: Clinical and Experimental Research* 9:331–333. Another study, by Merikangas et al., presented evidence that depression and alcoholism are not usually manifestations of the same disorder (*Archives of General Psychiatry* 42:367–372).

Based on recent studies, perhaps one third of alcoholics suffer from anxiety disorders that appear to precede the onset of heavy drinking. These anxiety disorders—particularly agoraphobia in women and social phobias in men—may be a cause or consequence of drinking, but at least sometimes appear to be the cause. (For a review of anxiety disorders and alcohol abuse see the September 1984 issue of *The Journal of Studies on Alcohol* and the January 1985 issue of *The Journal of Clinical Psychiatry.*)

p. 38
The study challenging the view that most people with Laennec's cirrhosis are alcoholics was reported by Coates et al. in *Clinical Investigative Medicine* 9:26–32, 1986. This is still a controversial position. (See review by Leevy et al. in the *Acta Medica Scandinavia* supplement 703:675–679.)

pp. 39–40
Alcohol has long been condemned as a destroyer of brain cells, but the evidence has not been substantial, in part because of what

appears to be reversible atrophy in CT scan studies, plus differences of opinion among neuropathologists about whether gross or microscopic lesions of the brain are found in alcoholics more often than in nonalcoholics (except in the case of Wernicke–Korsakoff patients and a certain small percentage of patients with cerebellar disease). A 1987 study in the *British Medical Journal* 294:534–536, using advanced technology, counted brain cells in alcoholics who came to autopsy and indeed found that alcoholic patients had fewer cortical neurons in the frontal cortex than did nonalcoholics. If confirmed by further studies, this would support observations by Courville and others, many years ago, attributing cortical atrophy of the frontal lobes to alcohol. Whether the reported loss of neurons in the present paper has functional significance is not certain, but with the new techniques for counting brain cells, clinical–pathological correlations will now be possible.

The difficulties in interpreting computed tomography studies in alcoholics is illustrated by two 1986 papers, both by Jacobson (*British Journal of Addiction* 81:661–669 and *Psychological Medicine* 16:547–559). Women alcoholics revealed CT scan abnormalities after a markedly shorter drinking history and at a lower estimated peak alcohol consumption than male alcoholics, but the data were confounded by differences in the nonalcoholic control subjects, with males having larger CT brain ventricles than healthy females of the same age. The studies also suggested that CT scan abnormalities might disappear with abstinence, particularly in female alcoholics.

p. 39

A 1986 review by Schaeffer and Parsons (*Alcohol* 3:175–179) supported the assumption that in alcoholics, the greater and more frequent the alcohol ingestion, the more the brain's functioning is disrupted, at least as reflected in psychological test performance. The article stresses the importance of anxiety in test performance. Mild psychological test impairment is sometimes reported in social drinkers, but this review suggests anxiety may be the cause.

p. 40
The role of thiamine deficiency in the Wernicke–Korsakoff syndrome is reviewed by Yellowlees in *The Medical Journal of Australia* 145:216–219.

p. 42
The tentative conclusions about alcohol and blood pressure are based on articles by Arkwright et al. (*Circulation* 66:60–66) and Maheswaran et al. (*Clinical Science* 72:70).

p. 42
The heretofore unreported association between stroke and alcohol consumption was studied by Gill et al. in the *New England Journal of Medicine* 315:1041–1046.

p. 44
"Lunar Caustic" appeared in the *Paris Review* in the Winter–Spring issue of 1963.

p. 49
Vaillant's valuable work on the natural history of alcoholism was published by Harvard University Press in 1983. For more information on natural history, see *Longitudinal Research in Alcoholism,* edited by Goodwin, Van Dusen, and Mednick (Kluwer-Nijhoff, 1984).

pp. 50–52
In the past few years the literature on women and alcohol has become voluminous. Kalant edited a huge book on the subject (*Alcohol and Drug Problems in Women,* Plenum Press, 1980). Other excellent works include a 1984 book edited by Wilsnack and Beckman called *Alcohol Problems in Women* (Guilford Press). Taylor and St. P. Pierre provide an up-to-date (1986) review of women and alcohol research in the *Journal of Drug Issues* 16:621–636.

p. 51
Support for the view that most women alcoholics abuse other drugs is contained in an article by Bissell and Skorina in the *Journal of the American Medical Association* 357:2939–2944.

pp. 54–55
The two Seattle pediatricians were K. L. Jones and D. W. Smith. The paper that made fetal alcohol syndrome famous appeared in *Lancet* 2:999 (1973). Since then there have been hundreds of reports on FAS, including some that are critical of the concept. The late Henry Rossett performed one of the largest studies of FAS and failed to find the syndrome in most offspring of drinking women (*Alcohol and the Fetus,* Oxford University Press, 1984). Abel and Sokol took a dire view of the syndrome in *Drug and Alcohol Dependence* 19:51–70.

Proponents of FAS rarely acknowledge negative evidence on the subject. In fact, the great majority of alcoholic women do not have children with overt physical or mental abnormalities. When abnormalities occur, the explanation may be malnourishment rather than alcohol per se. In fact, there are studies dating back more than half a century indicating that alcohol intoxication during pregnancy does not produce adverse effects in the offspring. One study, conducted in the Galton Eugenics Laboratory in 1910, reported that children of alcoholics were *more* healthy than children of nonalcoholics! (K. Pearson and E. M. Elderton, *Galton Eugenics Laboratory Memoirs,* issues 10, 14 and 17). In another study, 30,000 mice were carried through numerous generations in a continuously intoxicated state. The offspring were normal. In yet another study, poultry experts judged the merits of a large number of chickens. Some had been conceived by drunken progenitors, some by sober progenitors. The offspring of the drunken chickens were rated as superior, being larger and having finer combs and feathers.

The poultry study was described in a book by Walton Smith and Ferdinand Hellwig titled *Liquor: The Servant of Man* (Little, Brown, 1939). The authors maintained that liquor probably improves the human race by selectively destroying "inferior" eggs and sperm. The theory is a dubious one, but the results were unequivocal. Other studies indicate that the fetus is tremendously resilient. Even where the mother is in a state of starvation during pregnancy, there does not appear to be long-term adverse effects on the offspring. (See C. A. Smith, "The Effect of Wartime Starvation in Holland Upon Pregnancy and Its Product," *American Journal of Obstetrics and Gynecology* 55:599–606, and Z. Stein et al., "Nutrition and Mental Performance: Prenatal Exposure Seems Not Related to Mental Performance at Age 19," *Science* 178:708–713.)

p. 60
Richard L. Solomon has published widely on the opponent-process theory of acquired motivation, which roughly parallels speculations about the addictive cycle. He describes the theory in the August 1980 issue of *American Psychologist.*

p. 62
Other sources used for this chapter, as well as the chapter on treatment, were *Drug and Alcohol Abuse* by Marc Schuckit (Plenum Press, 1984); *Medical and Social Aspects of Alcohol Abuse,* edited by Tabakoff, Sutker, and Randall (Plenum Press, 1983); and *Diagnosis and Treatment of Alcoholism,* second edition, edited by Jack Mendelson and Nancy Mello (McGraw-Hill, 1985). The three volumes are all by highly esteemed authorities and can be recommended.

CHAPTER THREE

pp. 63–64
In the last century, if constitutional weakness, Lamarckian inheritance, or the devil (or all three) were not blamed for a man's alcoholism, the presumed precipitant, if environment was given any due, most likely was masturbation. There was a time—unbelievably less than a century ago—when masturbation was considered by distinguished physicians and educated laity alike as the leading cause of nervousness, insanity, and other problems now called psychiatric. "Masturbatory disease" was the admitting diagnosis of large numbers of patients in mental asylums. It

would be surprising if at that time masturbation had not been considered a contributory cause to alcoholism.

p. 64
The historical background, by and large, came from a review by Warner and Rossett in the *Journal of Alcohol Studies* 36:1395–1420.

p. 64
The Hebb quote came from the *Textbook of Psychology*, published by W. B. Saunders Co. in 1958 (chapter 6).

pp. 64–65
For studies confirming alcoholism as a family disease, see a review by Nancy Cotton in the *Journal of Studies on Alcohol* 40:89–116.

p. 65
There have been no nationwide attempts to determine the prevalence of alcoholism in the United States. The studies by Cahalan and associates, mentioned earlier, focused on alcohol problems. Being sociologists, the investigators rejected the whole concept of alcoholism.

p. 65
The formula for calculating alcoholism rates based on cirrhosis was formulated by Jellinek and Keller in a 1952 paper in the *Quarterly Journal of Studies of Alcohol* 13:49–59. Based on cirrhosis data, alcoholism affected about 5 percent of American men—very close to the figure derived from European studies. The latter studies are reviewed in a paper by D. W. Goodwin in the *Archives of General Psychiatry* 25:545–549. The German study was reviewed in the same paper.

p. 66
Some will argue that hospitalized alcoholics do not split fifty-fifty with regard to a family history of alcoholism. A huge study of thousands of sailors showed the fifty-fifty break, but the ratio obviously will be changed depending on the sample. The 1933

study was by K. Pohlisch and was published in German. The data were reviewed in the paper by Goodwin cited above. The issue of specificity to addictions came out in a study of Vietnam War veterans (Goodwin et al., *Archives of General Psychiatry* 32:230–233) in which some used heroin and some drank alcohol (almost all smoked marijuana). Alcoholism in the family predicted alcohol abuse in Vietnam and afterwards but did not predict heroin use in Vietnam, supporting the view that addictions tend to run pure to type in families.

On the other hand, animal studies support nonspecific addiction proneness. Nichols found that "whatever genetic endowment that predisposes rats to morphine relapse also appears to predispose them to alcoholism relapse as well. The phenomenon suggests a common mechanism of addiction to opiates and to alcohol" (*Annals of the New York Academy of Sciences* 197:60–65).

pp. 67–68
The fact that higher rates of alcoholism are determined through personal interviews with the relatives of alcoholics was found by Rimmer and Chambers and published in the *American Journal of Ortho-psychiatry* 39:760–768.

p. 68
Other sources of error in family studies have been reviewed by Goodwin in the *Journal of Studies of Alcohol* 42:156–162. Researchers generally agree that interviews are superior to questionnaires in gathering such data, but even structured interviews in which a form is used have drawbacks. One never knows whether the questions were asked the same way. One rarely knows the experience, training, biases of the interviewers (who range from moonlighting housewives to venerable academics). One never knows whether the subject understood the question, listened, lied, or had a faulty memory.

Regarding memory, the question "How well do people remember life changes?" was raised in a study by Jenkins et al. in the *Archives of General Psychiatry* 36:379–384. Nearly 500 normal subjects were interviewed on two occasions separated by nine months. At the second examination, subjects were asked to report events during the six-month period preceding the first examination. Nine months later, subjects "forgot" more than one third of the events reported originally, including such items as being in an automobile accident or being the victim of a crime.

Also, one rarely knows how much weight to give an answer. If a man was scolded once by a fundamentalist wife for having a beer at a ballgame, do you score "family complains about drinking" positive? If he had one blackout at age nineteen, do you score him positive for blackouts? If he *thinks* he had a blackout but isn't sure, what then? It requires judgment by the interviewer, and interviewers differ widely in judgment and other ways.

Finally, few family history studies (maybe none) deal with the problem of conflicting stories. The patient says his father was a drunk, the patient's sister says the father drank a lot but never had problems, and the patient's paternal uncle said the father drank no more than the uncle himself, which was modest indeed. How does one reconcile conflicting stories? They are inevitable, given differing opinions of what constitutes a drinking problem, plus the fact that, among mobile Americans, many of us know little about our relatives. Still, no one mentions conflicts. One suspects the investigator's bias probably determines which version is accepted.

pp. 69–71
John Barleycorn can be found in a 1982 collection of Jack London works published by the Viking Press.

p. 71
Numerous studies support the observation that familial alcoholism begins at an early age and is particularly severe. A 1985 review by Goodwin in the *Archives of General Psychiatry* 42:171–174 lists the studies and makes the point that there seems almost no exception to the early-onset and severity correlations.

p. 72
Cook and Winokur reported that familial alcoholics more often had relatives with antisocial personality disorder than did non-familial alcoholics (*Journal of Nervous and Mental Disease* 173:-175–177). The Danish adoption studies failed to find this

relationship (chapter 5). Hesselbrock and associates also reported an association between familial alcoholism and antisocial personality (*Neuro-psychopharmacological Biological Psychiatry* 6:607–614). Penick and colleagues, in a large VA sample, found that familial alcoholics had higher rates of additional psychiatric disorders than did nonfamilial alcoholics—a finding at variance with the adoption studies, among others (*Journal of Studies on Alcohol* 48:136–146). It is not clear whether the additional psychiatric disorders, such as depression and anxiety conditions, were a consequence of the drinking or preceded the drinking. The same is true of the association of alcoholism with antisocial behavior. Lower-class alcoholics when drunk notoriously have trouble with the law, but are not necessarily criminal when sober.

The "which comes first" problem has led to the concept of primary and secondary alcoholism. The concept is entirely chronological. Secondary alcoholics have psychiatric illnesses preceding the alcoholism; primary alcoholics develop their alcoholism "out of the blue" without preceding psychiatric illnesses. Whether the preceding illnesses are actually causes of the alcoholism is not specified, but often assumed. As noted frequently in this book, the author holds the tentative opinion that primary and familial alcoholism are largely identical, but, as seen, this is not everyone's view. Powell and associates believe the primary–secondary distinction may have clinical relevance and certainly should be explored further (*Journal of Clinical Psychiatry* 48:98–101).

p. 72
An association of alcoholism and a childhood history of hyperactivity has been reported by a number of investigators (reviewed by Tarter et al. in the *Journal of Studies on Alcohol* 46:259–261). The evidence that familial alcoholics are particularly likely to have such a history is somewhat contradictory, but at present it points in that direction.

p. 72
Oscar Parsons and colleagues found that familial alcoholics showed more deficits on certain psychological tests than did nonfamilial alcoholics, and there was evidence the deficits existed

before heavy drinking began. The Parsons group reported that alcoholism and a positive family history of alcoholism had independent, additive deleterious effects on cognitive–perceptual functioning (Scheffer, Parsons, and Yohman in *Alcoholism: Clinical and Experimental Research* 4:347–351).

pp. 72–73
Frances et al. report that "not only does familial alcoholism predict a poorer treatment outcome, but also the greater the number of first-degree relatives, the worse the prognosis . . . just as the familial forms of depression and schizophrenia tend to have a poor prognosis, we find that familial alcoholism is also more chronic and difficult to treat" (*American Journal of Psychiatry* 141:1469–1471.)

p. 73
Wayne London at the Brattleboro Retreat in Brattleboro, Vermont, has published numerous reports on the association of left-handedness and familial alcoholism. His latest paper is in press in the *Journal of Nervous and Mental Diseases.* Other supporting papers are by Smith and Chyatte in the *Journal of Studies of Alcohol* 44:553–555 and by Bakan in *Perceptual Motor Skills* 36: 514.

p. 73
It may have nothing to do with left-handedness, but Alterman and Tarter reported that familial alcoholics more often have a history of head injury than do nonfamilial alcoholics (*Journal of Studies on Alcohol* 46:256–259). The investigators speculated that possibly antedating conditions, such as hyperactivity or a conduct disorder, believed to be more prevalent in children of alcoholics, or perhaps adult antisocial personality disorder, augment the risk of coincidental head trauma. In any case, the history of head trauma may be a significant factor contributing to poor treatment outcome.

p. 73
The biological alcoholic parents in the adoption studies are more often the fathers than the mothers. This is why little is known

about the mothers' drinking habits during pregnancy. Adoption records do not usually provide such information.

p. 77
Mayfield's study appeared in *Alcoholism and Affective Disorders,* pp. 99–108, edited by Goodwin and Erickson (Spectrum Publications, 1979).

p. 78
The possible relationship between alcoholism and affective disorders, including manic-depressive disease, is examined comprehensively in *Alcoholism and Affective Disorders* (S. P. Medical and Scientific Books, 1979), edited by Goodwin and Erickson.

p. 79
The relationship between anxiety disorders and alcoholism is referenced on page 218.

p. 82
For a review of psychiatric disorders in general, and their relationship to drinking in particular, see *Psychiatric Diagnosis* (Oxford University Press, 1988) by Goodwin and Guze.

CHAPTER FOUR

p. 83
A recent unpublished study by the National Academy of Sciences indicates that genes more than environment are responsible for money growing on family trees. Two thousand pairs of twins were studied. Identical twins, born of the same egg, showed a greater similarity in earnings than fraternal twins, born of two eggs. There was some evidence that parents treat identical twins more equally than they treat fraternal twins, but this did not seem to explain the difference.

p. 84
The Strömgren quote is from an article in the *Proceedings* of the Second International Congress of Human Genetics, pp. 1757–1761, Excerpta Medica Foundation, 1962.

pp. 85–86

The Medawar quote is from *Hope of Progress,* pp. 59–60 (Anchor Books, 1973).

pp. 86–87

Regarding environmental triggers: Although a genetic factor can be clearly implicated in the cause of a disease (particularly if it follows a precise Mendelian mode of inheritance, such as Huntington's Chorea), failure to identify a genetic factor does not necessarily mean it does not exist. An example of this is glucose-6-phosphate dehydrogenase deficiency disease, where a particular genetic makeup is required, but where the illness does not become symptomatic unless triggered by certain drugs. Another example is porphyria, a genetic disease that may not be recognized until symptoms are precipitated by barbiturates. In the case of alcoholism, an environmental trigger obviously is necessary: alcohol.

pp. 90–91

In the Danish adoption studies described in chapter 5, we had good assurance that the father listed in the adoption records was indeed the father. In the case of illegitimate births, Danish law requires that the mother identify the child's father. There is no stigma or penalty attached, and presumably most of the mothers identified the father correctly.

Drunkenness and illegitimacy, by the way, have long been associated, and not always with unhappy results, as demonstrated by this story from Montaigne's sixteenth-century essay "On Drunkenness":

> A widow of chaste repute, feeling the first inklings of pregnancy, told her neighbors that she might think she was with child if she had a husband; but when from day to day her suspicion grew into evident certainty, she went so far as to authorize the priest to announce from the pulpit that, if any man should avow himself privy to the deed, she promised to pardon and, if he approved, marry him. A young laborer in her service, emboldened by this proclamation, declared that he had found her one holiday so much under the influence of wine, so fast asleep, and in so indecent a posture by her fireside, that he had been able to ravish without awakening her. They are still living as man and wife.

p. 93
Pollock et al. reported the alpha finding in the *Archives of General Psychiatry* 40:857–861.

pp. 93–94
The categories test finding was described by Gabrielli and Mednick in the *Journal of Nervous and Mental Disorders* 171:444–447. While at Seattle, Marc Schuckit found that sons of alcoholics had higher acetaldehyde levels than sons of nonalcoholics. He repeated his findings after moving to San Diego, and later the findings were reported by two other groups. However, some studies have been negative. One reason for the contradictions is the tricky technology involved in measuring acetaldehyde. There are people in the field who hardly speak to each other because they can't agree on measuring this elusive chemical.

p. 94
Marc Schuckit is the father of the high-risk study in alcoholism. (The idea really originated with Fini Schulsinger and Sarnoff Mednick, who started high-risk studies of schizophrenics in the early sixties.) Schuckit's laboratory has been extremely productive. Among other findings, he has reported that, compared to controls, sons of alcoholics become less intoxicated when drinking, have a greater behavioral tolerance for alcohol, and have greater muscular relaxation after drinking. His evidence for innate tolerance is impressive. Recently he has reported that sons of alcoholics are even tolerant to the hormonal effects of alcohol. Sons of alcoholics have lower levels of prolactin (a pituitary hormone) and cortisol (an adrenal hormone) after drinking than do sons of nonalcoholics (*American Journal of Psychiatry* 144:854–859).

pp. 94–95
Begleiter and associates were the first to report that sons of alcoholics had diminished amplitude of the P300 wave (*Science* 225:1493–1495). This was later confirmed by six other groups studying nonintoxicated subjects. In two studies of intoxicated subjects, abnormalities in P300 were found in one study and not

in the other. In a 1987 report, John Polich and Floyd Bloom at the Scripps Clinic failed to replicate Begleiter's finding (*Alcohol* 4:301–305).

One problem with P300 studies, as well as with most high-risk studies, is that subjects usually are recruited in their late adolescence when, based on all evidence, the sons of alcoholics have already begun heavy drinking. Heavy drinkers are always excluded from the studies on the grounds that (1) the drinking may confound the results and (2) it may not be ethical to include people with possible drinking problems in a study that involves giving alcohol. As a result, many high-risk studies systematically may exclude from their samples precisely those sons of alcoholics who later will become alcoholic and include those who will not. As frequently noted in this book, familial alcoholism typically begins early in life, often by the mid-teens. This is a catch-22 problem in high-risk studies that seems insoluble.

p. 95
It is interesting what is *not* found in high-risk studies. Blood-alcohol concentration, for example, is the same in sons of alcoholics and controls. No personality differences have been reported. With regard to psychological test data, there is much conflict in the literature, with Schuckit's group unable to find any differences between sons of alcoholics and controls in a recent study (unpublished data).

p. 95
A 1987 paper in the *Journal of Clinical Investigation* (79:1039–1043) reported finding a potentially interesting marker for alcoholism in a high-risk study. Nondrinking sons of alcoholics more often had a particular enzyme abnormality than did controls. The enzyme was transketolase. Thiamine depends on transketolase for its activity, and thiamine is strongly implicated in the Wernicke–Korsakoff syndrome, one of the most serious neurological complications of alcoholism. Attempts to reproduce the findings should be made because of its potential importance.

p. 96
The longitudinal studies are reviewed in George Vaillant's book *The Natural History of Alcoholism* (Harvard University Press, 1983).

p. 96
Hyperactivity in sons of alcoholics is reviewed by Tarter et al. in the *Journal of Studies of Alcohol* 46:259–261.

p. 96
Ethnic differences in alcoholism rates are discussed and documented in Dwight Heath's chapter on "Sociocultural Variants in Alcoholism," in the *Encyclopedic Handbook of Alcoholism,* edited by E.M. Pattison and E. Kaufman (Gardner Press, 1982).

p. 96
Alcohol and the Writer: an American Epidemic by D. W. Goodwin (Andrews McMeel, 1988) proposes that a genuine epidemic of alcoholism among American writers occurred in the first half or two thirds of the twentieth century. It is very difficult, indeed, to think of many American writers during that period who were *not* alcoholic. The alcoholic Nobel Prize laureates were Sinclair Lewis, Eugene O'Neill, William Faulkner, and Ernest Hemingway. John Steinbeck probably would have been called alcoholic by most observers today. Saul Bellow won the Nobel Prize and is not an alcoholic, but Bellow is Jewish and Jews have a low rate of alcoholism regardless of their occupations. St. Louis-born T. S. Eliot, was not an alcoholic, but confessed that he could not write poetry unless he first had a few drinks. The role of alcohol in the creative process is interesting, but particularly fascinating is the question of why alcoholism reached truly epidemic proportions among well-known writers in America during the twentieth century. There were alcoholic writers in other places and other times, but never in such large numbers. The cause of the epidemic remains a mystery.

CHAPTER FIVE

p. 97
Galton is the founder of the Eugenics movement. The quotation is from his book *History of Twins, as a Criterion of the Relative Powers of Nature and Nurture* (written in 1875 and reprinted by Cambridge University Press in 1911).

p. 100
Another advantage of doing the study in Denmark was the opportunity it gave to work with Dr. Fini Schulsinger, then director of the psychiatry department of the Kommunehospitalet in Copenhagen and codirector with Sarnoff Mednick of the Psykologisk Institut. Dr. Schulsinger has much experience in adoption studies. His friendship has been much valued. Dr. Schulsinger is now Secretary General of the World Health Organization, but still very active in research.

Dr. Mednick, another valued friend, provided help and advice throughout the study.

Other Danish friends and colleagues who worked on the adoption studies included Drs. Leif Hermansen, Joachim Knop, and Niels Møller, all fully trained psychiatrists. They did the interviews in Danish and translated them into impeccable English. My American mentors and colleagues on the adoption study were Drs. Samuel B. Guze and George Winokur. Dr. Guze is now vice chancellor for medical affairs at Washington University in St. Louis, as well as the head of its psychiatry department. Dr. Winokur is chairman of the department of psychiatry at the University of Iowa.

The Danish studies would not have been possible without the help and generosity of Drs. David Rosenthal, Seymour Kety, and Paul Wender. Together with Drs. Schulsinger and Mednick, they organized the first Danish adoption study in the late 1960s. They were interested in schizophrenia, another illness that runs in families. Their cooperation is warmly appreciated.

p. 99
The Danish adoption studies were sponsored by the National Institute of Alcohol Abuse and Alcoholism under grants AA-00256 and AA-47325, and Research Scientist Development Award AA-

47325. The work was published in detail in four issues of the *Archives of General Psychiatry,* in 1973 (28:238–243), in 1974 (31:-164–169), and in 1977 (34:751–755 and 34:1005–1009).

p. 100
Data pertaining to education in the Danish adoption studies can be misleading, given the educational system in Denmark. Academic schooling is often followed by various forms of vocational or other training that is equivalent to high school or college education in the United States. The data presented conform closely to academic educational levels in the population from which the sample was drawn. Social class estimates were based on a rating scale used by other Danish adoption studies (K. Svalastoga, *Prestige, Class and Mobility,* Gyldendahl, 1959).

p. 107
The observation about class bias in adoptions was made by Fini Schulsinger, codirector of Denmark's Psychological Institute (personal communication).

p. 114
Winokur and associates proposed that, in certain families, women are depressed and their male relatives are either alcoholic or sociopathic. They refer to this cluster as depressive spectrum disease. They first described it in the *British Journal of Psychiatry* in 1975, 127:75–77, and elaborated on the concept in *Alcoholism and Affective Disorders,* edited by Goodwin and Erickson (S. P. Medical and Scientific Books, 1979).

p. 115
Bohman's first adoption paper appeared in the *Archives of General Psychiatry* in 1978 (35:269–276). Two 1981 papers by Bohman and Cloninger also appeared in the *Archives of General Psychiatry* (38:861–868; 38:965–969). The latter two papers introduced the concept of two types of alcoholism. Cloninger stressed the importance of environmental factors in a paper in the *Journal of Psychiatric Treatment and Evaluation* (5:487–496) in 1983. Both Bohman and Cloninger have published widely, either separately or together. A recent paper by Cloninger (*Science,* April 24,

1987) ambitiously reviews clinical, genetic, psychopharmacological, developmental, personality, and learning studies in order to produce a unified theory for the development of alcoholism. The author's wide range of learning in the fields is most evident from the paper.

I am grateful to a personal communication from Dr. Bohman citing studies supporting the representativeness of his sample. He pointed out that there were big differences between the Danes and Swedes in their attitudes toward alcohol. Since there was strong agreement between his 1978 paper and the Danish adoption papers, it was not clear what differences in attitude he was referring to. A paper by Kaij on possible biases in the Temperance Board records (*Social Psychiatry* 4:216–218) found that there was little difference between alcohol abusers who were officially registered and those who were not, except that the registered ones were more often psychotic.

p. 120
The Korean/Vietnam study was by Branchey et al. and appeared in *Alcoholism: Clinical and Experimental Research* 8:572–575.

pp. 120–21
There are two types of people: those who believe there are two types of people and those who do not. Cloninger and Bohman obviously belong to the latter group.

In the *Proceedings* of the Second Malmo Symposium on Alcohol in November 1984, Bohman et al. reported that Type 2 (male-limited) alcoholics had lower activity of monoamine oxidase (MAO) in their platelets than did Type 1 alcoholics or normals. The samples were small but the results intriguing. MAO in platelets is believed to reflect MAO activity in the brain. MAO degrades serotonin, among other amine neurotransmitters, and, assuming Type 2 alcoholism is synonymous with familial alcoholism, the finding has interesting implications for the serotonin hypothesis developed in chapter 2.

pp. 121–24
Cadoret's first adoption study appeared in 1978 in the *British Journal of Psychiatry* 132:252–258. A second adoption study ap-

peared in the same journal in 1980 (37:561–563), and a third in 1985 (42:161–167). The *Archives of General Psychiatry* published adoption papers by Cadoret et al. in 1985 (42:161–167) and 1986 (43:1131–1136). In 1987 another Cadoret et al. adoption study appeared, this time in the *Journal of Studies on Alcohol* (48:1–8). In addition to this remarkable series of papers, Cadoret did an excellent critique of adoption studies, published in *Psychiatric Developments* in 1986 (1:45–64).

pp. 124–25
The Roe paper appeared in Memoirs of the Section on Alcohol Studies, Yale University, *Quarterly Journal of Studies on Alcohol,* No. 3, 1945.

p. 125
The Åmark monograph was Supplement 70 in the *Acta Psychiatrica Neurologica Scandinavia,* published in 1951.

p. 126
The half-sibling paper by Schuckit, Goodwin, and Winokur appeared in the *American Journal of Psychiatry* 128:122–126.

p. 126
An ongoing series of studies at the Institute for Behavioral Genetics at the University of Colorado in Boulder involves adoption and twin registers. Gabrielli and Plomin recently reported findings from a study of 346 adult Colorado twins, nontwin sibling pairs, and pairs of unrelated adoptees reared together (*Journal of Studies on Alcohol* 46:24–31). The results suggested that identical twins shared a more common environment than did fraternal twins and did not support the contention that being raised in the same family contributes substantially to similarity in characteristics of drinking behavior. Apparently there were no alcoholics in the sample and the drinking behavior could be considered normal. Nevertheless, the results somewhat contradict those reported by Cadoret and the Cloninger–Bohman group, suggesting that environmental factors influence drinking behavior.

p. 126
Two excellent reviews of adoption and other genetic studies of alcoholism were written by Peele (*Journal of Studies on Alcohol* 47:63–73) and Thacker et al. (*Alcoholism: Clinical and Experimental Research* 8:375–383).

pp. 126–27
Kaij's study is described in his book *Alcoholism in Twins,* published by Almquist and Wiksell in 1960.

p. 127
The study by Partanen et al. is described in their book *Inheritance of Drinking Behavior,* published by the Finnish Foundation for Alcohol Studies in 1966.

pp. 128–29
The Hrubec and Omenn study was published in *Alcoholism: Clinical and Experimental Research* 5:207–215, 1981.

p. 129
The Maudsley studies are described in the chapter by Murray, Clifford, and Gurling in volume 1 of *Recent Developments in Alcoholism,* edited by Marc Galanter (Plenum Press, 1983).

p. 130
The National Merit Scholarship twin study was published by Loehlin in the *Annals of the New York Academy of Science* in 1972 (197:117–120).

p. 130
The Cederlof et al. paper was published in the *Acta Medica Scandinavia* 202:1–128.

p. 130
The Jonsson and Nilsson paper appeared in *Nordic Hygiene Tidskr* 49:21–25.

p. 130
The Italian study was by Conturio and Chiarelli and appeared in *Heredity* 17:347–359.

p. 131
The Kapprio et al. study appeared in *The Finnish Twin Registry: Baseline Characteristics, Section 2* (University of Helsinki Press, 1978).

p. 131
Richard P. Swinson has an excellent review of marker studies in volume 1 of *Recent Developments of Alcoholism,* edited by Marc Galanter (Plenum Press, 1983).

p. 132
Cruz-Coke and Varela first published the colorblindness finding in *Lancet* 2: 1282.

p. 133
The success by T.-K. Li and associates in producing "alcoholic" rats is described in a paper in *Alcohol and Alcoholism,* supplement 1, pp. 91–96, 1987.

p. 135
In 1972, Peter Wolff first observed that Orientals flushed and complained of nausea after drinking (*Science* 175:449–450). It is astonishing that no one would report this interesting phenomenon until 1972. Since then, there have appeared numerous papers on Oriental flushing. The enzymes involved in producing the flush seem to have been identified. Meanwhile, at Kansas University, investigators have been successful in blocking the flush with a combination of antihistamines (Miller et al., *Journal of Studies of Alcohol,* in press).

CHAPTER SIX

pp. 136–40
Many inherited diseases are confined primarily to specific ethnic or racial groups. Sickle-cell anemia, for example, occurs chiefly

in blacks and cystic fibrosis mainly in whites. Six of the most devastating genetic illnesses are found predominantly in Jews. These include Tay-Sachs, Niemann-Pick, and Gaucher's diseases, all caused by metabolic abnormalities. To learn about the causes and possible cures of these diseases, a National Foundation for Jewish Genetic Diseases has been established. A National Foundation for Jewish Genetic *Non*diseases might also be established, with the primary focus on alcoholism.

p. 142
The idea that a certain fixed percentage of alcohol users will become alcohol abusers, and the ratio applies to all users in every society, is contradicted by the facts. For example, almost all Jewish adults use alcohol on some occasions and yet the percentage of Jewish abusers is small. Among Southern Baptists, relatively few use alcohol but of those who do, a sizable proportion are abusers. In the general population, however, the ratio of abusers to users is probably about one in ten or one in twelve.

p. 144
The first rat study showing a nurture effect was by Komura et al. and appeared in the *Quarterly Journal of Studies on Alcohol* 33:-494–503. Similar findings were reported by Randall and Lester in *Science* 189:149–151.

Randall and Lester also gave evidence that "mothering" includes intrauterine influences (*Nature* 255:147–148). They first showed that fertilized eggs transplanted from mice with a high preference for alcohol to a uterus of nondrinking mice continued to show high preference. This tended to support a genetic basis for alcohol preference in mice, but there was a hitch: Fertilized ova from nondrinkers were transplanted to drinker mothers, which also led to increased alcohol intake by the offspring. In short, the data supported both a genetic and environmental explanation for alcohol preference (in mice).

p. 145
H. J. Clinebell lists four ways in which alcoholics make bad parents in an article in *Alcoholism,* edited by Ronald Cantanzaro (Thomas, 1968).

p. 145
There is not much literature on teetotalers. A paper by Goodwin et al. in *Comprehensive Psychiatry* (10:209–215) indicated that very few individuals become teetotalers because of the "negative example" of having an alcoholic parent. A paper by Lucero et al. supports this conclusion (*Quarterly Journal of Studies of Alcohol* 32:183–185).

p. 146
For more about O'Neill's drinking habit, see D. W. Goodwin's article in the *Journal of the American Medical Association* 216:99–104.

p. 146
Sources consulted about Irish drinking customs included an essay by John Henry Raleigh in *O'Neill: A Collection of Critical Essays,* edited by John Gassner (Prentice-Hall, 1964), *The Irish Countryman* by S. T. Kimball (Macmillan, 1937), *Irish Land and Irish Liberty* by M. J. F. McCarthy (Scott, 1911), and an excellent review of these and other works by Robert F. Bales in *Society, Culture and Drinking Patterns* (Wiley, 1962).

p. 147
The either/or attitude toward drinking and sex has old roots. In ancient times people believed that nations of moderate drinkers or abstainers indulged in sexual excess. Thus, in Plato's *Laws,* Magillus of Lacedaemon, who extols his nation's abstinence from wine, is told by the Athenian: "An Athenian in self-defense might at once retaliate by pointing to the looseness of the women in your country!"

p. 148
Evidence that depression does not automatically lead to drinking is included in a paper by Cassidy et al. in the *Journal of the American Medical Association* 164:1535–1546.

pp. 148–49
I am grateful to Dr. Gilman Ostrander for permitting me to use excerpts from his letter. M. K. Bacon (*Quarterly Journal of Stud-*

ies on Alcohol 35:863–876) analyzed more than one hundred preliterate societies and found evidence supporting Dr. Ostrander's theory. Bacon found that "pressures toward achievement in childhood" led to frequent drunkenness in adulthood and that "indulgence in infancy and tolerance of dependent behavior in adulthood" was associated with a low degree of drunkenness—Ostrander's theory in a nutshell.

pp. 150–53
The number of COAs in the United States was taken from a 1986 article in the *Journal of the American Medical Association* by MacDonald and Blume, "Children of Alcoholics" (140:750–754). Since Dr. MacDonald was director of the Federal Mental Health Program, the statistic can be taken seriously. The article also is a good review of the many studies linking specific mental and behavioral problems to the trauma of growing up with an alcoholic parent.

p. 151
Califano was quoted in *Psychiatric News* (June 19, 1987).

pp. 151–52
Illustrative of the large body of anecdotal literature on COAs is an article by Cutter and Cutter in the *Journal of Studies on Alcohol* 48:29–32. Interviewing members of Al-Anon, the authors found that "spirituality" bound the group together as much as anything, demonstrating how unrepresentative such samples can be.

p. 152
In the *British Journal of Psychiatry* (149:584–591), Beardslee and Vaillant compared 176 men who had grown up with an alcoholic parent with 230 men without such exposure. They found more impairment in the sons of alcoholics, but most of the impairment occurred in those subjects who developed alcoholism. In short, alcoholism ran in families but having an alcoholic parent, even when raised by the parent, did not necessarily increase one's risk of developing other psychiatric disorders (precisely the finding of the Danish studies described in chapter 5).

pp. 152–53
Few studies compare children of alcoholics with children of nonalcoholics. One study was published by Jacob and Leonard in 1986 in the *Journal of Studies on Alcohol* 47:373–380. Comparing children of alcoholics and children of depressed parents with children of parents who were neither alcoholic nor depressed, they found no differences between the children of alcoholics and the children of depressed people. Both groups had somewhat more behavior problems than did the control children, but not many more.

In the *Journal of Studies on Alcohol* 48:265–268, Parker and Harford compared children of "heavy-drinking parents" with children without such exposure. Their main finding was that the children of heavy drinkers were more often heavy drinkers than were controls and also came from a lower socioeconomic status. They reported no evidence that the children of alcoholics were more likely to have other psychiatric disorders.

p. 153
A 1987 paper by Workman-Daniels and Hesselbrock failed to find, as have most other papers, that children of alcoholics were intellectually inferior to children of nonalcoholics (*Journal of Studies on Alcohol* 48:187–193).

p. 153
The 1987 telephone survey polled 2,000 New York State residents sixteen and older. The following was found:

1. Of the total questioned, 16.6 percent said one or both parents had an alcoholism problem.
2. Among COAs and non-COAs alike, only 5 percent identified COAs as being a high-risk group for alcoholism.
3. Among males, COAs drank more than those who were not COAs.
4. Among COAs, those who did not see themselves at risk drank three times as much and got drunk nine times as often as did those who were aware of the risk.

Source: Division of Alcoholism and Alcohol Abuse, 194 Washington Avenue, Albany, NY 12210.

p. 154
Jellinek's remarks can be found in a chapter called "Heredity of the Alcoholic" in *Alcohol, Science and Society* (Greenwood Press, 1945). Roe describes her adoption study in the next chapter.

p. 156
In 1985 the U.S. Department of Health and Human Services published and widely distributed a little book called *Alcoholism: An Inherited Disease.* (No question mark.) The idea that alcoholism is inherited clearly has received wide acceptance (a big change from ten years ago) and now is official doctrine. The Government may be a little premature.

CHAPTER SEVEN

p. 158
The wonderful quote by Roueché comes from his book *Alcohol* (Grove Press, 1960).

pp. 165–66
The Lewis Thomas quote is from *The Medusa and the Snail* (Viking Press, 1979).

p. 168
D.F. Horrobin discusses the cyclic effects of alcohol on prostaglandins in *Medical Hypothesis* 6:929–942.

p. 170
The chemical name for Antabuse is disulfiram.

pp. 170–71
G.L.S. Pawan wanted to see if exercise, caffeine, fructose, or glucose would speed up the rate of alcohol metabolism (*Proceedings of the Biochemical Society, Biochemical Journal* 106:19–21). Only fructose worked, and the quantities required were too large to be practical.

p. 171
There may be one problem in adding vitamins to alcoholic beverages. There is evidence that alcohol blocks the absorption of thiamine (B_1) and probably other vitamins. This may explain why individuals may have a vitamin deficiency if they drink alcohol despite being well nourished. A variant of thiamine called allithiamine has been studied in Japan, and its absorption is not inhibited by alcohol. It is worth looking into. Chemists at the Stroh Brewery Company in Detroit, Michigan, have found that the addition of thiamine to beer does not alter its taste (personal communication from Morton C. Meilgaard, vice-president of research).

p. 172
In one study, antidrug films were shown in one community but not in another. The community where the films were shown ended up with a higher prevalence of drug use than did the community where they were not shown. Again, many other factors may have been involved, and the films themselves may not have been very good.

p. 175
A 1986 review by Gurling (*Psychiatric Developments* 4:289–309) makes an educated guess about where genes for alcoholism might be located on specific chromosomes. Meanwhile, in addition to Huntington's chorea, a paper in the February 26, 1987, issue of *Nature* by Egleland and associates announced what might be another triumph for gene mapping. By cutting DNA with restriction enzymes (the RFLP approach), the researchers were able to visualize distinct sites on the chromosomes and compare variations in DNA sequences. Studying a large Amish family in Pennsylvania, they discovered a strong correlation between the presence of manic-depressive illness and two "marker" genes at the tip of the short arm of chromosome 11. The two markers—the gene that encodes insulin and a so-called oncogene—border a region where DNA sequences vary considerably in length in this Amish pedigree. According to the study's statistician, the odds were greater than 1,000 to one that a linkage exists between this gene locus and the presence of manic-depressive

illness. He speculated that the "manic-depressive gene" most likely was autosomal dominant, meaning an individual need inherit only one gene from either parent to develop the disorder. The gene, he noted, was not completely "penetrant." Only about 63 percent of the Amish with the manic-depressive gene actually developed manic-depressive symptoms.

Unfortunately, the same issue of *Nature* contained reports by two other teams of investigators reporting failure in linking genes with manic-depressive disorder. The media seemed to focus on the positive report and ignore the negative ones, which is not unusual. Meanwhile, studying variations in gene sequences represents a very direct approach to finding a genetic basis for alcoholism, if there is one.

CHAPTER EIGHT

p. 177

There is much talk about the need for treatment evaluation studies, but, in fact, treatment has been studied ad nauseam (if not always well). Hundreds of evaluations have been published in the last forty years. They all come to remarkably similar conclusions:

1. There is no evidence that one treatment is superior to another.
2. There is no evidence that treatment is superior to no treatment (although most investigators believe treatment *does* work, and say so).
3. Characteristics of the patients, rather than treatment, seem the most important predictors of outcome. Patients with stable marital and occupational status and higher socioeconomic status have better outcomes.
4. Alcoholics treated as outpatients seem to do as well as, and perhaps slightly better than, those treated as inpatients.

Reviewing this enormous literature has been helped considerably by several excellent reviews. Recommended are reviews by Voegtlin and Lemere (*Quarterly Journal of Studies of Alcohol* 2:717–750); Hill and Blane (*Quarterly Journal of Studies of Alcohol* 28:76–95); Emrick (*Quarterly Journal of Studies of Alcohol* 35:523); Baekland, Lundwall, and Kissin (*Research Advances in*

Alcohol and Drug Problems, Volume 2, edited by Israel, Wiley, 1975); Costello (*Encyclopedic Handbook of Alcoholism,* edited by Pattison and Kaufman, Gardner Press, 1982).

p. 178
A classical study indicating that outpatients fare as well as inpatients was performed by Edwards and Guthrie (*Lancet* 1:555–561). Supporting the point is a review by Wanburg et al. in the *International Journal of Mental Health* 3:160–170.

p. 178
The paper by Miller appeared in the July 1986 issue of *American Psychologist.*

p. 178
The Helen Annis summary of the 1980 Institute of Medicine findings are from a *Science* article (236:20–22).

p. 178
Psychiatrists are notoriously reluctant to see alcoholics. Many view alcoholics as people who miss appointments, call at all hours, and test the psychiatrist's patience by refusing to get better. Even when treating a patient with a drinking problem, some psychiatrists interpret the drinking as symptomatic of some other condition and ignore it.

Alcoholism is often ignored by other physicians as well. A survey of a large medical ward revealed that 25 percent of the male patients had a serious drinking problem, which may have contributed to their illness, but the hospital charts rarely mentioned drinking (Barchha et al., "The Prevalence of Alcoholism Among General Hospital Ward Patients," *American Journal of Psychiatry* 125:681–684). Another study found that before some physicians suspect alcoholism the alcoholic has to be dirty and unshaven (Chafetz in the *American Journal of Psychiatry* 124:1674–1679). Until a few years ago many general hospitals in the country would not admit patients with a diagnosis of alcoholism. Alcoholics often take a dim view of doctors. Since the feeling is often reciprocated, this attitude is understandable.

Some alcoholics refuse any treatment that does not come from other alcoholics. "How can anyone help an alcoholic who has not been one?" goes the reasoning. The same principle could apply to any condition. How can anyone treat schizophrenia or diabetes who has not been schizophrenic or diabetic? Some feel they can, and do, with some success.

The real issue concerns not who treats alcoholics but who treats them best. Until social workers or recovering alcoholics can show they get better results than medical doctors, or vice versa, professional chauvinism seems ill-advised. (Alcoholics prefer the term "recovering" to "recovered" on the AA-grounded belief that one *never* recovers from alcoholism. "Reformed alcoholic" is blasphemous.)

pp. 179–83
Any reader who thinks the review of psychological therapies of alcoholism in this book is overcritical is urged to read "A Review of Psychologically Oriented Treatment of Alcoholism," by Emrick in the *Quarterly Journal of Studies on Alcohol* 36:88–108. Emrick reviewed 384 studies and found no connection between treatment methods and results.

p. 181
The dependency hypothesis cuts both ways: Alcohol has been attributed to underdependency and to overdependency. The problem with both views is that the recognition of dependency needs rests on inference, not on direct observation. Inference is often presented as if it were directly observable. Just as "innate heat" was used in Aristotle's time to explain body temperature, "dependency needs" have been adduced to explain alcoholism.

p. 181
One reason psychiatrists in private practice may find alcoholism difficult to treat is that they often do not recognize it when it exists. A study of alcoholism treatment programs conducted by the Joint Information Service of the American Psychiatric Association and the National Association for Mental Health found that some psychiatrists do not know about drinking habits of their

patients because the patient who does not volunteer such information may not be asked. Several times the following dialogue was reported:

Staff member of alcoholism treatment program to patient: Have you previously had psychiatric treatment?

Patient: Yes, from a psychiatrist in private practice.

Staff member: What did he tell you about your drinking?

Patient: Nothing. I never told him about it, and he never asked me. (From Glasscote et al., *The Treatment of Alcoholism,* American Psychiatric Association, 1967.)

p. 181
The theory that alcoholism represents a form of self-destruction is eloquently presented in Karl Menninger's *Man Against Himself* (Harcourt, Brace, 1938).

p. 182
For a description of the way transactional analysis can be applied to the treatment of alcoholism, see *Games Alcoholics Play* by Claude Steiner (Grove Press, 1971).

p. 183
Medawar's comment on nonspecific aspects of psychotherapy was made in his book *The Hope of Progress,* pp. 59–60 (Anchor Books, 1973).

pp. 183–84
Alcoholics Anonymous (AA) has been assiduous in maintaining its anonymity. In protecting this indisputable right, it has resisted the importunings of scientists to investigate it. From the standpoint of AA, it has probably been a good thing: Like blacks and homosexuals, alcoholics are grist for the social scientists' mill. At the same time it has held back the advancement of knowledge.

pp. 184–88
For a good review of behavior therapy see the article by Miller and Barlow in the *Journal of Nervous and Mental Disease* 157:10–20. Voegtlin has an article on conditioning therapy in the *American Journal of Medical Sciences* 199:802–809, and another in the

Quarterly Journal of Studies on Alcohol 2:505–511. Aversion conditioning is described by Lund and Vogler in *Behavioral Research and Therapy* 8:313–314.

pp. 187–88
The case histories by William James can be found in *Principles of Psychology* (Holt, 1890).

p. 188
A new drug for anxiety called buspirone (BuSpar) was released in early 1987 and, unlike other sedative and antianxiety drugs, does not appear to interact with alcohol. In other words, based on studies, people can take the drug and drink without addictive effects from the combination. Studies cannot always be trusted, and it would be clearly unwise to advise patients to take the drug and go ahead and drink. But since many will anyway, it may be reassuring to know that probably taking buspirone and drinking can be done with relative safety.

p. 189
Fawcett's study was published in the *Archives of General Psychiatry* 44:248–256. Earlier studies indicating that lithium might be helpful for some alcoholics were published by Kline et al. in the *American Journal of Medical Science* 268:15–20, and by Merry et al. in *Lancet* 2:481–482 (1976).

pp. 190–91
A large number of side effects have been attributed to Antabuse, including rashes and psychotic reactions. These occur so infrequently that it is not possible to know whether they are caused by the drug. Clinicians inclined toward skepticism sometimes wonder whether side effects attributed by a patient to Antabuse may in fact be motivated by desire to stop taking the drug and start drinking again. In combination with alcohol, of course, there is no question that Antabuse produces serious effects.

p. 194
Even suggesting that a small percentage of alcoholics return to "normal drinking" for long periods is anathema for many recov-

ering alcoholics. Nevertheless, such evidence exists, including a 1987 paper by Nordström and Berglund in the *Journal of Studies of Alcohol* 48:95–103.

p. 195
The evidence that most relapses begin within the first six months following treatment can be found in an article by Skoloda et al. in the *Journal of Studies on Alcohol* 36:365–379 and a study by Davies et al. in the *Quarterly Journal of Studies on Alcohol* 17:-485–502.

p. 195
Evidence that internists have better success with alcoholics than do psychiatrists can be found in *Outpatient Treatment of Alcoholism* by Gerard and Sanger (University of Toronto Press, 1966). Since 1966, the situation may have changed and the author of this book, a psychiatrist, sincerely hopes so.

p. 198
Evidence that the CAGE test correctly identifies most alcoholics can be found in a 1982 *Lancet* article by Brown et al. (1:325–328).

Index

About the Author

DONALD W. GOODWIN, M.D., is Professor and Chairman of the Department of Psychiatry at the University of Kansas Medical Center in Kansas City. An international authority on alcoholism—and the first researcher to establish a link between heredity and alcoholism—he has written many books and several hundred scientific publications on the subject. Before entering medical school he was a columnist for *The Los Angeles Times* syndicate.